Rebuilding Fukushima

Five years after the one of the worst nuclear accidents in history, Fukushima now only occasionally headlines national and international media. However, the disaster is far from over, as evidenced by a hundred thousand people from Fukushima still in the state of evacuation, rising levels of radiation in streams and rivers, and failing attempts to control the leakage of radioactive materials at the Fukushima Daiichi Nuclear Power Plant. Despite these dismal conditions, efforts to recover and rebuild livelihoods in the afflicted regions of Fukushima did start immediately after the outset of the accident.

Rebuilding Fukushima gives an account of how citizens, local governments, and businesses responded to and coped with the crisis of Fukushima. It addresses principles to guide reconstruction and international policy environments in which the current disaster is situated. It explores how reconstruction is articulated and experienced at different spatial scales, ranging from individuals to communities and municipalities, and details recovery efforts, achievements, and challenges in the realms of public transportation, agriculture and food production, manufacturing industries, retail sectors, and renewable-energy industries. The book also critically investigates the nature of the current reconstruction policy schemes, and seeks to articulate what may be required in order to achieve more sustainable and equitable (re)development in afflicted regions and other nuclear host regions.

Drawing on extensive fieldwork and local surveys, this volume is one of the first books in English that captures the knowledge and insights of native Japanese social scientists who dealt with the complexities of nuclear disaster on a day-to-day basis. It will be of great interest to students and scholars of disaster-management studies and nuclear policy.

Mitsuo Yamakawa is Professor of Economic Geography at Teikyo University and Extraordinary Professor of the Fukushima Future Center for Regional Revitalization (FURE) at Fukushima University.

Daisaku Yamamoto is Associate Professor of Geography and Asian Studies at Colgate University, Hamilton, New York. His recent work focuses on community resilience, regional inequality, and uneven development.

Routledge Studies in Hazards, Disaster Risk and Climate Change

Series Editor: Ilan Kelman, Reader in Risk, Resilience and Global Health at the Institute for Risk and Disaster Reduction (IRDR) and the Institute for Global Health (IGH), University College London (UCL).

This series provides a forum for original and vibrant research. It offers contributions from each of these communities as well as innovative titles that examine the links between hazards, disasters and climate change, to bring these schools of thought closer together. This series promotes interdisciplinary scholarly work that is empirically and theoretically informed, with titles reflecting the wealth of research being undertaken in these diverse and exciting fields.

Published:

Cultures and Disasters
Understanding cultural framings in disaster risk reduction
Edited by Fred Krüger, Greg Bankoff, Terry Cannon, Benedikt Orlowski and E. Lisa F. Schipper

Recovery from Disasters
Ian Davis and David Alexander

Men, Masculinities and Disaster
Edited by Elaine Enarson and Bob Pease

Unravelling the Fukushima Disaster
Edited by Mitsuo Yamakawa and Daisaku Yamamoto

Rebuilding Fukushima
Edited by Mitsuo Yamakawa and Daisaku Yamamoto

Climate Hazard Crises in Asian Societies and Environments
Edited by Troy Sternberg

Rebuilding Fukushima

Edited by Mitsuo Yamakawa and
Daisaku Yamamoto

Routledge
Taylor & Francis Group
LONDON AND NEW YORK

First published 2017
by Routledge
2 Park Square, Milton Park, Abingdon, Oxon OX14 4RN

and by Routledge
711 Third Avenue, New York, NY 10017

Routledge is an imprint of the Taylor & Francis Group, an informa business

British Library Cataloguing in Publication Data
A catalogue record for this book is available from the British Library

Library of Congress Cataloging in Publication Data
Names: Yamakawa, Mitsuo, 1947-editor. | Yamamoto, Daisaku, editor.
Title: Rebuilding Fukushima / edited by Mitsuo Yamakawa and Daisaku Yamamoto.
Description: Abingdon, Oxon ; New York, NY : Routledge, 2017. | Includes index.
Identifiers: LCCN 2016036626| ISBN 9781138193796 (hardback) | ISBN 9781315639147 (ebook)
Subjects: LCSH: Fukushima Nuclear Disaster, Japan, 2011. | Radioactive pollution–Japan–Fukushima-ken. | Radioactive waste sites–Cleanup–Japan–Fukushima-ken. | Environmental disasters–Japan–Fukushima-ken. | Fukushima-ken (Japan)–Environmental conditions.
Classification: LCC TK1365.J3 R44 2017 | DDC 363.17/9970952117–dc23
LC record available at https://lccn.loc.gov/2016036626

ISBN: 978-1-138-19379-6 (hbk)
ISBN: 978-1-315-63914-7 (ebk)

Typeset in Times New Roman
by Sunrise Setting Ltd, Brixham, UK

MIX
Paper from responsible sources
FSC
www.fsc.org
FSC® C013056

Printed and bound in Great Britain by
TJ International Ltd, Padstow, Cornwall

Contents

Figures

Tables

Contributors

Toshio Hatsuzawa is Professor of the Faculty of Human Development and Culture at Fukushima University and Director of Fukushima Future Center For Regional Revitalization (FURE). His research interest includes industrial location and regional economies with particular emphasis on the study of local industrial promotion. His recent work includes "Nuclear power station disaster and local industries" (*Regional Economic Studies*, 2012; in Japanese) and "Problems of the industrial revival in the areas affected by the East Japan Great Earthquake Disaster" (*Textbook Studies of Disaster Revival Supports*, Hassosha, 2014; in Japanese).

Hideki Ishii is Project Associate Professor at the Fukushima Future Center For Regional Revitalization (FURE), Fukushima University. At the FURE, he works for the Division of Food and Agricultural Recovery. His specialty is landscape planning and cultivating experiments.

Satoru Mimura is Specially Appointed Professor of the Fukushima Future Center for Regional Revitalization (FURE) at Fukushima University. He is in charge of disaster awareness and international relations. He has a professional career in the Japan International Cooperation Agency and the Ministry of Environment in Japan, specializing in global environment and disaster management. He has also engaged in international dialogues on sustainable development while working with the World Bank on the project *Learning from Mega Disasters—Lessons from the Great East Japan Earthquake.*

Katsumi Nakai is Professor of the Faculty of Administration and Social Sciences at Fukushima University and the President of Fukushima University, Japan. His main research areas include legal science, public administration, and environmental law. He is a member of the Japan Public Law Association and the Japan Association for Environmental Law and Policy.

Takehito Noda is Assistant Professor of Environmental Sociology at the College of Policy Science, Ritsumeikan University, Osaka, Japan. His recent work includes "The meaning of noneconomic activities in a community business: Water resources utilized by tourism in Harie Village, Takashima City, Shiga" (*Journal of Environmental Sociology*, 2014; in Japanese) and "The important cultural landscape selection system as a means for maintaining the territorial integrity of a rural community: A case study of the community of Harie in Takashima City, Shiga Prefecture, Japan" (*Asian Rural Sociology*, 2014).

Takashi Norito is Project Associate Professor of Agricultural Economics and Geography at the Faculty of Economics and Business Administration, Fukushima University. His research interests include food systems, the food industry cluster, and regional development. His current work focuses on rebuilding local food systems in Fukushima, aiming at the industrial recovery of food and agriculture. His recent publications include "Structural features of the East Asian food systems and dynamics: Implications from a case study of develop-and-import scheme of Umeboshi" (*Geographical Review of Japan Series B*, 2012).

Yoshio Ohira is Assistant Professor at Hosei University. His PhD is in economics, and his research fields include environmental economics, industrial organization, and energy economics. His current research focuses on renewable-energy policies and electric-power generation in Japan. His recent publications include "A study of renewable energy policy in Fukushima for regional revival" (*Journal of Public Utility Economics*, 2013; in Japanese).

Akihiko Sato is Associate Professor in the Takasaki City University of Economics. He specializes in regional sociology with a particular focus on the mutual relationship between local autonomy and residents. His current research examines the structure of the issues surrounding the nuclear-accident evacuees in Fukushima, adopting the KJ method and using participant observation.

Masayuki Seto is Research Fellow of Geography at Fukushima University and a Post-Doctoral Research Fellow of Physical Geography at Kyung Hee University, Korea. He holds a DSc in geomorphology and is a member of the program committee of the Japan Geoscience Union. He is a contributing author to *Weathering: Types, Processes and Effects* (edited by M. J. Colon, Nova Science Publishers, 2011). His recent research interests include the modeling of disasters and their revival processes.

Akira Takagi is Associate Professor of Geography at Kumamotogakuen University, and has a PhD in geography. His current research focuses on the industrial revival after the 3.11 Disaster in Kawauchi Village, Fukushima, in comparison to the reconstruction experience after the volcanic eruptions in Miyake Island in 2000. His recent publications include "Comparison of the disasters and spatio-temporal scale" (*Abstracts and Proceedings, The 9th Korea-China-Japan Joint Conference on Geography*, 2014).

Mitsuo Yamakawa is Professor of Economic Geography at Teikyo University and Extraordinary Professor of Fukushima Future Center for Regional Revitalization (FURE) at Fukushima University. He has a PhD in Science from the University of Tokyo. He is also a Regular Member of the Science Council of Japan. His recent books include *Economic Geography on Revitalization from*

Fukushima Nuclear Disaster (Sakurai-shoten, 2013) and *Japanese Economy and Regional Structure* (as editor, Hara-shobou, 2014). He is the project leader on *Establishing Academic Framework of Earthquake Disaster Reconstruction Experiencing Great East Japan Earthquake*, funded by Grant-in-Aid Scientific Research (Category S) of Japan.

Daisaku Yamamoto is Associate Professor of Geography and Asian Studies at Colgate University, Hamilton, New York. He holds a PhD in geography from the University of Minnesota. His recent work focuses on community resilience, regional inequality, and uneven development. He is currently working on a project to examine the socio-economic effects of nuclear decommissioning on local communities in the United States and Japan.

Itsuki Yoshida is Associate Professor of the Faculty of Economics and Business Administration at Fukushima University, and has a PhD in urban science (Tokyo Metropolitan University). He has studied local-transport policies and planning, especially in rural areas of Japan, and was a member of the committee in the Ministry of Land, Infrastructure and Transport that established the Transport Basic Law in Japan. He is a co-editor of *Local Transport Planning for Citizens' Life* (Gakugei Publishing; in Japanese), which received the 34th Transportation Book Award from Transportation News Corporation.

Preface and acknowledgments

The earthquake on March 11, 2011, and the subsequent tsunami and nuclear accident took thousands of lives and forever changed the lives of tens of thousands of people who lived along the coast of the Tohoku region of Japan. Immediately after the earthquake and tsunami, the areas swept by their mighty and ruthless force resembled the landscape after the air raids of World War II that burned and leveled the cities of Japan: buildings destroyed without a trace and washed-up wreckage dotting the scene. Five years after the disaster, signs of recovery and reconstruction are evident in many of the devastated areas. However, in the towns near the Fukushima Daiichi Nuclear Power Plant (NPP) of the Tokyo Electric Power Company (TEPCO), you can still see buildings destroyed by the earthquake and tsunami, as if time came to a full stop on March 11. One cannot help feeling a sense of profound human futility in the face of a nuclear accident releasing a significant amount of radioactive materials across the landscape.

Indeed, while the damage from the earthquake and tsunami was enormous, it is environmental contamination by radioactive materials that has been the greatest source of affliction for many residents of Fukushima Prefecture. This is especially the case for inhabitants of the communities around the nuclear power plant who were ordered to evacuate, but it is also the case for those living outside of mandatory evacuation zones, yet still within higher-than-normal radiation areas. These individuals have had their families torn apart, their livelihoods suspended or altered, their health severely weakened, and their children's education interrupted. Many of the evacuated areas have been gradually labeled as "safe to return" after government-led decontamination projects; yet, the process of returning has been decidedly slow. Farmers and businesses still suffer from tangible damage from and stigma associated with radioactive contamination.

The human casualties, environmental damage, and socio-economic distress all represent what we have lost in this historic disaster. Yet, they are not all of what we lost. The earthquake, tsunami, and subsequent nuclear power accident, and the poor responses that followed, brought down heavy criticism and condemnation not only upon politicians, the government–corporate nexus, and the media, but also upon the academic community, and even scientific knowledge itself. These feelings of eroding faith in authority, experts, and science were encapsulated in and exacerbated further by the plethora of new phrases and terms that filled the media after March 11. Prominent among these terms are discourses of "the unexpected," "no immediate effects," and "interim limits." For example, the notion that a nuclear disaster is "unexpected" means that it is assumed to be beyond the range of the possible. It was on the foundation of this assumption—and also a flagrant over-confidence in the superiority of Japanese manufacturing and engineering—that the paradigm of nuclear power plant safety was erected. "No immediate effects" also carries in its train a whole host of troublesome baggage. This term emerges out of a politically motivated attempt to quell the anger and agitation of victims and the general public, but it can only be uttered through the willful disregard of substantial scientific evidence concerning the actual state of air radiation-dose levels in the afflicted areas and the health effects of radiation exposure. Simply relying on data from the International Commission on Radiological Protection (ICRP)—a group whose voice is among the choir chanting the refrain of nuclear safety—to assess low-dose radiation exposure's effects on children is by no means sufficient to dispel anxiety for parents and the public. Finally, the phrase "interim" has been used as a means of altering the limits for radiation in food for a certain but undefined period of time, and it is this declaration of an interim of exception that has been the source of much disbelief among victims and the public, leading to the issue of stigmatization, or reputational damage (*fuhyo higai*).

The appearance of these three paradigmatic examples of nuclear-disaster discourse and procedure are exemplarily illustrative of the way in which the formerly paired phrases in Japanese, "*anzen*" and "*anshin*" ("safety" and "reassurance"), have been clearly divided such that, even when objective criteria are used to proclaim "safety," we are now no longer able to feel "reassured" by the claim. Even so, various government agencies, technocrats, and public officials have continued on various occasions, such as at public hearings and press conferences, to "ask for the understanding and cooperation of the citizens"—a uniquely Japanese bureaucratic expression that tactfully asserts a one-way flow of correct, expert knowledge and implicitly belittles the public for its ignorance. We must demystify the so-called expert knowledge and power associated with it. To do so inevitably demands a critical reflection on knowledge that we produce. In a country where the development of academic institutions has been tightly linked with the process of modernization and industrialization, such critical reflection and demystification of knowledge is not easily undertaken.

We would like to emphasize, nevertheless, that we still believe in the importance of the knowledge that we produce, in this case, for its practical usefulness in the rebuilding and revitalization of livelihoods of the people who have been torn away from their familiar environment and communities, rather than focusing exclusively on criticizing the developmental state and other power asymmetries. It is in this context that a team of researchers—headed by one of the editors (Yamakawa)—at Fukushima University decided to pursue a multi-year transdisciplinary project, supported by the JSPS KAKENHI Grant, on recovery, reconstruction, and redevelopment from the Great East Japan Earthquake Disaster—the term we use in this book to refer inclusively to the triple disasters in Japan on

March 11, 2011 of the earthquake, the tsunami, and the nuclear accident—and afterward. Many of the contributors to this book and another book that is being published concurrently (Yamakawa and Yamamoto 2016) are members of this research project. Other contributors come from various research institutions in and outside of Japan and have been conducting research on Fukushima over the past several years. Despite the wide range of institutional and disciplinary backgrounds of these authors, they are united by a shared concern and desire to contribute to the rebuilding and revitalization of the livelihoods of those who were affected by the historic disaster.

One of the editors (Yamakawa) was working as an economic geographer in the School of Economics and Management at Fukushima University, located in the prefectural capital city, at the time of disaster. As the only national university in the prefecture, it felt an urgent need to respond to multiple challenges on issues ranging from direct damage caused by the earthquake and tsunami to radioactive contamination, support for more than 100,000 evacuees, and assistance for local communities and industries. In April 2011, the university established the Fukushima Future Center for Regional Revitalization (FURE) as its outreach institute to address these issues. Yamakawa became the first director of the Center upon the recommendation of the president of the university, Osamu Nittono, and the vice president, Akira Watanabe. He would like to express his profound gratitude to those two individuals for providing such challenging, yet rewarding, opportunities.

Seiichi Chiaki, the head of the Center's office, and Yutaka Yamazaki, the associate head, provided enormous support in developing the Center's organizational architecture and hiring the staff researchers. The Center started out with four divisions: the Children and Youth Support Division, Reconstruction Planning Support Division, Energy Environment Division, and Planning and Coordination Division. The following members of the Center played pivotal roles in research activities for reconstruction and in the publication of this book: Toshio Hatsuzawa (director of the center); Hiroyasu Shioya, Katsuhiko Yamaguchi, Ryota Koyama and Tomotaka Mori (division heads); and Yasumichi Nakai, Fuminori Tamba, Naoaki Shibasaki, Kencho Wawatsu, Michio Sato, Noriko Yoshinaga, and Atsushi Igarashi.

Some of the chapters in this book are based on research supported by the five-year KAKENHI Grant-in-Aid for Scientific Research (Category S), 2013–2017, through the Japan Society for the Promotion of Science. Yamakawa applied for this grant in March 2013, before his retirement from Fukushima University, and the grant was subsequently awarded in June of that year. The proposal for the grant was drafted by the members of the FURE, including Yosuke Nakamura, Itsuki Yoshida, Akira Takagi and Akihiko Sato. The research project has been carried out by a number of professors and instructors of Fukushima University (unless otherwise noted), including Akira Takagi, Akihiko Sato, Toshio Hatsuzawa, Satoru Mimura, Hideki Ishii, Hiroshi Kainuma, Katsumi Nakai (current president of Fukushima University), Kenji Ohse, Yoshio Ohira, Kyo Kitayama, Noritsugu Fujimoto, Koichiro Matsuo (Teikyo University), and Kosei Yamada (Teikyo

University). Furthermore, doctoral researcher Masayuki Seto, hired as a project manager, oversaw and managed the project while also contributing to the research itself. Without his dedication, the KAKENHI project would not have progressed. We thank all the individuals who engaged in this research project, which provided an important basis for the book.

The members of the FURE and of the KAKENHI project have been advancing their research, around the keyword *shienchi* ("support knowledge"), through intensive and numerous interactions with local governments, organizations, and residents in disaster-afflicted areas of Fukushima. In particular, the research would not have been possible without the generous cooperation, wisdom, and patience of Yuko Endo (mayor of Kawauchi Village), Katsunobu Sakurai (mayor of Minami-Soma City), Tamotsu Baba (mayor of Namie Town), Norio Kanno (mayor of Iitate Village), officers of these municipalities, officers of the Prefectural divisions, specialized personnel of not-for-profit organizations, and above all those evacuees who have been forced to live away from their homes. It is our hope that the Center, the research project, and this book will help their efforts to rebuild their livelihoods in various ways, however small they be.

Yamakawa, as a council member of the Japan Society for the Promotion of Science, has been involved in a number of advocacy activities related to the recovery and reconstruction of nuclear disaster-afflicted regions since March, 2011. In particular, he played central roles in proposing the "Urgent recommendation on the development of inspection systems as a counter measure to the stigmatization of food and agriculture as the result of the nuclear disaster" (September 6, 2013) and "Recommendations on the reconstruction of livelihood and housing for long-term evacuees as the result of the TEPCO Fukushima Daiichi Nuclear Power Plant accident" (September 30, 2014). Ryota Koyama and Fuminori Tamba provided substantial assistance in developing these proposals.

The editors had known each other prior to the disaster and also met at the annual conference of the Japan Association of Economic Geographers in Tokushima, Japan, in the spring of 2012. However, the idea of publishing an English-language book (which turned out to be two books) did not arise until one of the editors (Yamamoto) and Noritsugu Fujimoto co-organized a series of sessions, titled *Fukushima: Three Years Later*, at the Annual Meeting of the Association of American Geographers in Tampa, Florida, in April 2014. The sessions offered opportunities for Japanese researchers working on the KAKENHI and other projects on Fukushima, and those researchers outside of Japan who have paid close attention to the nuclear disaster in Fukushima, to exchange knowledge and experiences. Many of this book's chapters were developed based on the presentations at the sessions. We thank all those who participated in the sessions as presenters, commentators, and audience.

We also thank Jay Bolthouse, who translated some of the manuscripts that were originally written in Japanese. His work went beyond mere translation of texts to become the translation of knowledge through his critical attention to the quality of arguments, logical flows, and data presentation. Furthermore, we thank Bill Meyer (Colgate University), who provided much help in revising many of the manuscripts and offered a number of useful comments, and Yamamoto's undergraduate students Angelica Greco and Samantha Trovillion, who also offered their assistance in editing and proofreading the manuscripts. Working with Melissa Heller, Julia Feikens, Mallory Hart, Jessica Li, and Madelin Horner on the research on nuclear decommissioning also informed this project in important ways.

We extend our appreciation to Colgate University, its Research Council, and several of its departments and programs that offered financial support for part of the project, including the AAG sessions and the hosting of the symposium on Fukushima in April 2014. We would like to acknowledge the members of the editorial and production staff at Routledge, including Faye Leerink, Priscilla Corbett, and Laura Johnson, for their professionalism, patience, and unfailing helpfulness, especially working with someone who had limited experience in publishing books in English.

Finally, we would like to express our greatest appreciation and deepest indebtedness to Ryoko Yamakawa and Yumiko Yamamoto for all their support and patience during the much-too-long process of the project leading up to this book.

Reference

Yamakawa, Mitsuo and Daisaku Yamamoto (eds). 2016. *Unravelling the Fukushima Disaster*. London: Routledge.

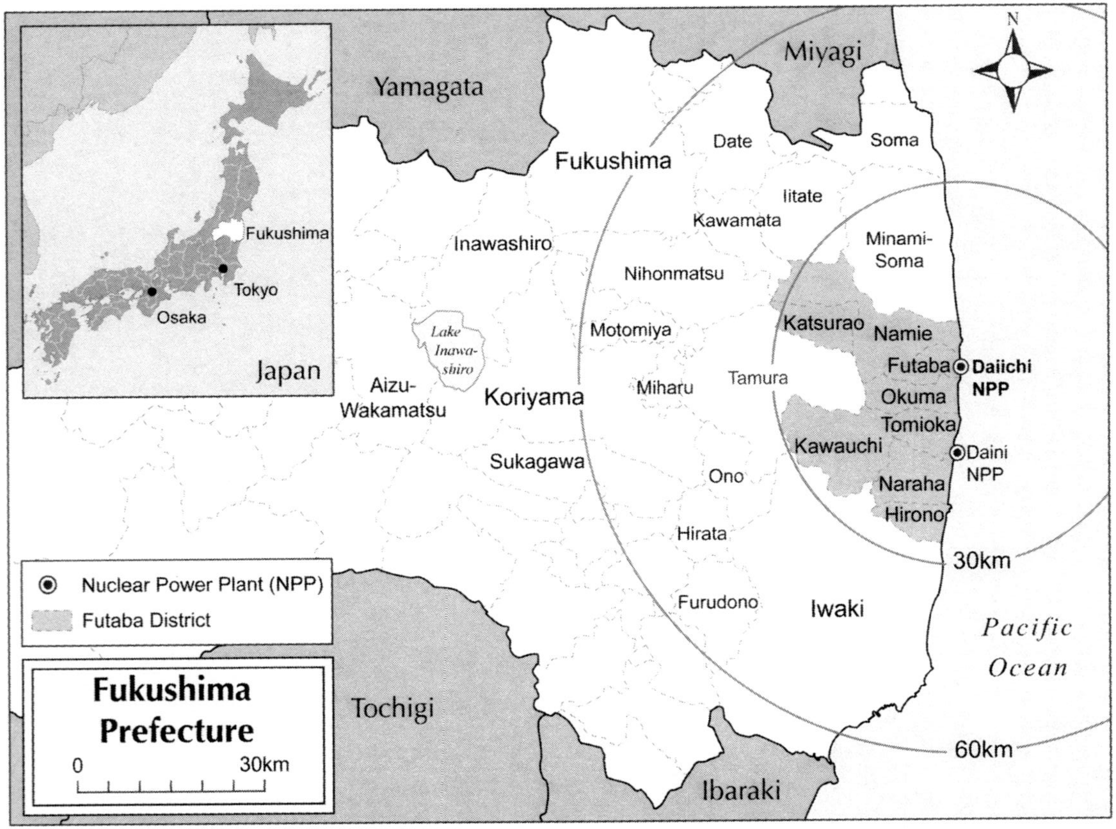

Map of Fukushima Prefecture.
Created by Masayuki Seto and Daisaku Yamamoto

Introduction

Mitsuo Yamakawa and Daisaku Yamamoto

In February 2016 a meeting was held in Fukushima City to discuss a proposal to establish a park to commemorate the reconstruction after the Great East Japan Earthquake Disaster in Fukushima Prefecture. The proposed site is symbolically located near the coastline of Futaba District and only 5 km from the Fukushima Daiichi Nuclear Power Plant. In a series of earlier preliminary meetings, members of the advisory council for the project—including present chapter author Yamakawa, who chaired the project—had urged that the proposed memorial park should give material form to two widely held sentiments: sorrow for the victims of the tsunami disaster and anger over the nuclear disaster. In a similar vein, one local resident serving as a representative on the committee expressed a widely felt duty to pass down to future generations the experiences of the tsunami, earthquake, and nuclear accident. Despite these urgings, in the six-page draft basic plan crafted by the prefectural office and presented at the February 2016 meeting, the phrase "nuclear accident" was conspicuous only for its complete absence. While the final proposal for the memorial park has yet to be released, the careful (and one must assume deliberate) excision of the phrase "nuclear accident" from the draft proposal is indicative of the current "atmosphere" (i.e. *kuki*) through which officials aim to tacitly cloud over critical conceptualizations of the disaster and its aftermath.

In the years after March 2011, a number of analyses of the Fukushima disaster appeared in Japan and beyond. A large number of publications in English focus on the processes that led up to the disaster, including the causes of the accident (Carpenter 2012; Hindmarsh 2013), the unfolding of the nuclear-reactor meltdown and hydrogen explosions (Lochbaum *et al.* 2014; Nadesan 2013; Willacy 2013), and the immediate reaction of the nation (Elliott 2012; Kingston 2012). Others provide analyses essentially so that the same accident won't happen again, discussing necessary political and institutional measures to prevent future disasters (Hindmarsh 2013; Kingston 2012; Samuels 2013; Aldrich 2011, 2013a, b) and how other countries should learn lessons from the Fukushima nuclear disaster (Elliott 2012; Lochbaum *et al.* 2014). More studies will certainly emerge over the next few years to advance both more sophisticated theoretical apparatuses and more refined analytical techniques. However, robust empirical analyses of the processes of recovery, reconstruction, and redevelopment of the nuclear disaster-afflicted areas remain all too thin, even in Japan, and we fear that interest in Fukushima may wane before we understand the actual extent and complexity of the multifaceted damage caused.

This book aims to keep the much-needed conversation on the Fukushima nuclear disaster alive and on our collective public agenda. In so doing we are strongly opposed to the overly optimistic portrayal of the situation in and around Fukushima (e.g. officials eager to claim "reconstruction complete," perhaps around the time of the 2020 Tokyo Olympics). At the same time, our aim is not to hastily conclude that Fukushima is so thoroughly damaged as to be beyond recovery; nor do we aim to lodge accusations against the national government or the Tokyo Electric Power Company (TEPCO) merely for the sake of accusation. Rather, our goal is to bring to light the specific damages and challenges resulting from the nuclear accident, to document both the successes and failures of rebuilding livelihoods, and to record concrete lessons that might eventually be used to develop a "Fukushima model" for nuclear disaster recovery.

The studies in this volume have been conducted by native Japanese researchers and thus reflect academic cultures that perhaps differ to some extent from Anglo-American academic cultures, especially in terms of their strong emphasis on empirical detail. Also reflected in each study is the fact that most of the authors have had to deal with the complexity of radioactive contamination on a day-to-day basis, since they were based in Fukushima University either during or after the nuclear disaster. Just like their fellow citizens and neighbors, these authors have been forced to pose to themselves and their families the difficult question of whether to remain in the area or to seek some form of "voluntary" removal. Many of the authors have also been put into a position, certainly not of their own choosing, to offer concrete policy recommendations amid a still unfolding and highly uncertain situation. In short, what each author has been forced to confront is the responsibility of a scholar embedded in a disaster-afflicted community. The following chapters, along with the concurrently published *Unravelling the Fukushima Disaster*, can thus be seen as records of their engagement with the disaster and early reconstruction efforts in Fukushima's vast "gray zones"—the large areas located in the complex and indeterminate geographical spectrum beyond the uninhabitable evacuated areas, but still within areas where above-normal radiation levels remain a concern. The situation in Fukushima remains fluid and uncertain, and some of the facts and data in this collection, despite our best efforts, quickly became dated; yet, we believe that this volume reflects well, and helps us remember, the evolving conditions in the afflicted communities of Fukushima over the past five years.

Overall directions for recovery and revitalization

Yamakawa and Nakai (Chapter 1) outline five principles to guide efforts to support victims and evacuees of the nuclear disaster. These principles cover a wide range of issues, including the restoration of confidence in government, compensation and monitoring schemes, community revitalization, forms of governance, and long-term and structural economic change. It should be emphasized that these guidelines have been informed by the authors' experiences in leading the Fukushima Future Center for Regional Revitalization (FURE), an outreach institute established at Fukushima University immediately after March 2011. Hence their propositions and recommendations sharply differ from those of national government agencies, and from those neatly derived from abstract theories, of social justice for example. Accordingly, while some might dismiss their proposals as too eclectic, unrealistic, or even self-contradictory, the fact is that their arguments are based on extensive surveys of refugees from Futaba District as well as interventions of the FURE staff researchers, and thus strongly reflect the voices of the disaster victims and refugees.

The magnitude of a disaster is determined in the interplay between the force of a natural phenomenon or human-induced accident and a society's internal capacities for resistance and response. The number of disasters occurring around the world, and their associated damages, have dramatically increased in recent years. Combined with increased economic integration and flows of people across national borders, there is a growing awareness that large-scale disasters are no longer confined to a country's internal affairs. Mimura (Chapter 2) describes the evolution of international frameworks for disaster reduction, highlighting the notion of "Build Back Better" as one of the most important new approaches to pre- and post-disaster measures. He suggests, however, that the Fukushima nuclear disaster is a serious test case for this approach—one that presents many difficulties when compared to most natural disasters.

Local governance, social organizations, and infrastructure

Yamakawa and Yamamoto (2016) argue that the damage from the Fukushima nuclear disaster can be conceptualized into three layers of issues that emerged successively, and accumulated one upon another over the course of the past five years. The three layers roughly correspond to three temporal stages of evacuation: immediate evacuation, transition into temporary housing, and settling into permanent housing. Sato (Chapter 3), in turn, shows that evacuees are facing multiple and interrelated problems that arise on three different spatial scales: personal/family, community, and local government. Drawing on case studies of two nuclear disaster-afflicted municipalities, Iitate Village and Tomioka Town, Sato highlights the inadequate functioning of local self-governance, especially the lack of any two-way dialogue, as one of the major factors that has hampered the reconciliation of problems across scales. Sato's analysis reveals that, despite the current national government's push to return evacuees to their hometowns, there seems to be deteriorating confidence in government, at both national and local levels, among the evacuees.

Noda (Chapter 4) argues that the issues facing nuclear disaster-afflicted localities are not simply a product of the geographical dispersion of former residents but also a result of the destruction of existing "life organizations" in the name of "rationalization" and "creative reconstruction" (akin to Naomi Klein's

"Shock Doctrine" argument). Life organization is a concept widely used in the Japanese sociological literature to reference social organizations that are intricately related to the daily lives of inhabitants of a local community. Using the case of Kawauchi Village, which experienced "whole-village evacuation" but became the first village to call for its residents to return after evacuation orders were lifted, Noda describes the process in which life organizations came under threat, but also how one hamlet in the village has attempted to maintain one of its few remaining life organizations—a localized water-supply system using stream water from a mountain in the area. The source of this water supply is within 20 km of the Fukushima Daiichi Nuclear Power Plant, and the residents have been advised by the village office not to use this water due to possible radiation contamination. Noda explores the communal logic behind resistance to abandoning this traditional water system, and suggests that such a seemingly trivial life organization may actually have significant social value and may play a pivotal role in supporting the rebuilding of post-disaster community life.

Securing the mobility of people is critical during a disaster, as well as during periods of prolonged evacuation in which evacuees must conduct their lives away from, but with occasional visits to, their homes. Yoshida (Chapter 5) examines the post-disaster response and transformation of public transportation in Minami-Soma City, a coastal municipality in Fukushima that has been significantly affected by both the earthquake/tsunami and the nuclear accident. In depicting the changing situation and need for mobility during the emergency, transition, and recovery periods, Yoshida foresees a growing difficulty in securing adequate local public transportation. The issue is that while some of the post-disaster responses and challenges regarding local public transportation stem from unordinary circumstances resulting from the massive disaster, many other issues are manifestations of pre-disaster socio-institutional conditions, including the declining population base and the lack of strategic plans for local public transportation integrated into a broader community-development planning framework. To put it another way, what the chapter demonstrates is that the quality of transportation policies in normal times has a decisive impact on the course of responses to a large-scale disaster.

Rebuilding local economies

One of the major effects of this nuclear disaster has been the radioactive contamination of agricultural fields and foods produced in Fukushima. The prefecture has been known as an agricultural prefecture, with the primary sector having a gross output of 147.6 billion yen in 2010. Of all agricultural products, rice makes up 34 per cent of sales, followed by vegetables (23 per cent), livestock (23 per cent), and fruit (13 per cent). After the nuclear accident, gross agricultural output declined from 129.0 billion yen to 102.5 billion, a drop of 21 per cent, due to the reduction of cultivated areas and a drop in the prices of crops due to stigmatization. It is in this context that Ishii (Chapter 6) addresses the challenges faced by the agricultural sector in post-nuclear disaster Fukushima. Ishii first explains the mechanisms whereby rice plants absorb cesium, a major radioactive substance released by the accident. He then discusses the importance of measurement and mapping of radioactivity in soils, the implementation of cultivation methods that reduce the transfer of radioactive substances to crops, and the development of radioactivity-monitoring systems for agricultural products. In particular, this chapter highlights the importance of linking cultivation-level strategies, including mapping and radioactive-absorption mitigation, with distribution-level strategies, such as agricultural-product inspection.

While Ishii focuses primarily on the issues that have arisen in agricultural fields and in crops as a result of the nuclear disaster, Norito (Chapter 7) examines the recovery of food systems—interconnected processes of food production, processing, distribution, and consumption around particular food commodities. The radioactive contamination and associated stigma (*fuhyo higai*: "reputational damage") seriously affected local food systems in Fukushima. However, through a case study of the Fukushima Soybean Project (FSP), Norito shows that although the system was damaged it did not completely break down, and that various key actors, including farmers, manufacturers, and cooperatives, are striving to overcome the challenges raised by the disaster. The chapter traces the sources of the system's resilience to a viable business model of support though purchase, an orchestrated synergy between consumer cooperatives and agricultural cooperatives, and the organizational culture of cooperation cultivated among the FSP members since the project's inception in 1998. The novelty of this study is that it provides us with a concrete understanding of socio-organizational mechanisms under which the recovery and possibly the strengthening of food systems might be possible after a major disaster.

Fukushima, due to its accessibility to Tokyo, has developed a strong manufacturing base over the past decades, especially in the areas of information and communication technologies, chemical engineering, and electronic parts and devices. Fukushima's manufacturing output dropped from 1,640 billion yen in 2010 to 1,330 billion in 2011. Hatsuzawa (Chapter 8) notes that although the initial impacts of the earthquake, tsunami, and nuclear accident on manufacturing industries in the Tohoku region have been reported in some journalistic and scholarly accounts in English, very few have followed their recovery process. Hatsuzawa's study, which focuses on the struggles for manufacturing recovery in the Haramachi ward of Minami-Soma City with a basis in multi-year surveys, therefore provides us with important empirical insights to shed light on this issue. One of the central findings of this study is a growing "recovery gap" between companies that are recovering and those that are not, and this can be best explained by the disaster accentuating the pre-disaster difference in business performance based on technological levels among firms.

Takagi and Seto (Chapter 9) take us back to Kawauchi Village, which was also the study site for Noda (Chapter 4). In this chapter, the authors focus on the efforts to restore and improve commercial functions and opportunities in the village, which was once evacuated entirely. Retailers in Kawauchi used to depend on the wholesalers and suppliers in larger towns of Futaba District, but because those towns, such as Tomioka Town, continue to be designated as mandatory evacuation zones, the previous supply chains cannot be restored. Nevertheless, there are signs of successful rebuilding of commercial services in the village, and this has been aided by such activities as the utilization of an existing but non-operational retail site for a new national-chain convenience store and collaboration with the Japanese Consumers' Co-operative Union. Because the rebuilding of commercial functions is integrally linked with the prospects of villagers' return, the authors argue, it is imperative to continue tackling this issue.

Ohira (Chapter 10) addresses the issues and challenges associated with renewable energy in Fukushima. Renewable-energy resources can potentially provide both energy and jobs in Fukushima. Nevertheless, there are several major problems that need to be overcome if renewable energy is to be promoted in Fukushima, including financing, human resources, and securing of land. Based on these conditions, this chapter outlines policy recommendations for the promotion of renewable energy in Japan and Fukushima. In particular, Ohira reveals the importance of cooperation by existing local industries and actors in promoting the renewable-energy sector, and that it is thus imperative for the people of Fukushima to become the main agents of renewable energy and to have rights and responsibilities regarding these operations.

In the final chapter (Chapter 11), we attempt to characterize the process of post-nuclear disaster reconstruction over the past five years. The chapter argues that this process is best characterized by its affinity to the logic of the developmental regulatory regime, or developmentalism, which has been the dominant institutional structure of postwar Japan. The system makes economic growth its main objective while allowing government intervention in the market in achieving the objective. These principles are tactfully held even during the post-disaster reconstruction period in the form of an emphasis on decontamination and return of evacuees, industrial projects in afflicted regions, and massive and coordinated capital investment to enhance safety measures in existing nuclear power plants, while leaving emergency evacuation plans behind. We also explore the implications of current policy orientations towards nuclear power for other localities that host nuclear power plants in Japan, and what may be required in order to achieve more sustainable and equitable (re)development of these localities. It is our view that only by continuing to speak about and maintaining a critical stance on this historic disaster and its aftermath can we draw useful lessons for Fukushima and elsewhere.

References

Aldrich, Daniel P. 2011. "Future Fission: Why Japan Won't Abandon Nuclear Power." *Global Asia* 6(2): 62–67.

Aldrich, Daniel P. 2013a. "A Normal Accident or a Sea-Change? Nuclear Host Communities Respond to the 3/11 Disaster." *Japanese Journal of Political Science* 14(2): 261–276.

Aldrich, Daniel P. 2013b. "Rethinking Civil Society–State Relations in Japan after the Fukushima Accident." *Polity* 45(2): 249–264.

Carpenter, Susan. 2012. *Japan's Nuclear Crisis*. New York: Palgrave MacMillan.
Elliott, David. 2012. *Fukushima*. New York: Palgrave Macmillan.
Hindmarsh, Richard. 2013. *Nuclear Disaster at Fukushima Daiichi*. New York: Routledge.
Kingston, Jeff. 2012. *Natural Disaster and Nuclear Crisis in Japan*. New York: Routledge.
Lochbaum, David, Edwin Lyman, Susan Q Stranahan, and The Union of Concerned Scientists. 2014. *Fukushima*. New York: The New Press.
Nadesan, Majia Holmer. 2013. *Fukushima and the Privatization of Risk*. New York: Palgrave Macmillan.
Samuels, Richard J. 2013. *3.11*. Ithaca, NY: Cornell University Press.
Willacy, Mark. 2013. *Fukushima*. Sydney: Macmillan.
Yamakawa, Mitsuo and Daisaku Yamamoto. 2016. "Refusing Facile Conclusions and Continuing to Tackle an Aggregating Disaster." In *Unravelling the Fukushima Disaster*, edited by Mitsuo Yamakawa and Daisaku Yamamoto, pp. 154–170. London: Routledge.

1 Five principles for the reconstruction of the nuclear disaster-afflicted areas

Mitsuo Yamakawa and Katsumi Nakai

Introduction

The earthquake and tsunami of March 11, 2011 caused massive destruction and took many precious lives. Despite the unprecedented magnitude of the disaster, the process of rebuilding started as soon as the ground stopped trembling. Damaged roads and ports have been rebuilt, new houses constructed, and people have started new lives. Yet it has become apparent that the nuclear disaster presents fundamentally different challenges. The nature of radioactive contamination makes it difficult, and in some cases impossible, even to start the process of recovery and reconstruction to make the land habitable again. The nuclear disaster has also been highly socially divisive. It has created divisions, for example, between generations in a household due to different perceptions of radiation exposure; between neighbors due to micro-scale differences in the levels of contamination; and between communities due to different evacuation-area designations.

We encountered this historic nuclear accident as faculty of Fukushima University. To say the very least, our lives have changed completely since March 11, 2011. We have both been deeply involved in the establishment and operation of the Fukushima Future Center for Regional Revitalization (FURE), a university-based research and outreach institute to support the recovery and reconstruction of the lives of people in Fukushima.[1] This chapter draws on our knowledge and experiences as directors of FURE in order to outline key principles for the reconstruction of the nuclear disaster-afflicted communities in Fukushima.

Background and basic conditions for reconstruction

In addition to the physical destruction of the land, health issues, and social divisions caused, we must first realize that a critical rift has formed between notions of safety ("*anzen*," or objective standards of protection) and confidence ("*anshin*," or subjective feeling of reassurance) as a result of the nuclear disaster. Prior to this disaster, objective safety standards were tightly tied to feelings of public confidence through the "nuclear safety myths" cleverly crafted by agents of the "nuclear village" (Samuels 2013) to instill confidence and safety in nuclear power in a country with a strong aversion to nuclear weapons.

These myths were created and reinforced through discourses propagating such notions as "severe accidents will never happen, given the types and levels of nuclear technology and safety measures that Japan has"; "atomic power for peace, nuclear weapons for war"; and "nuclear power is a 'quasi-domestic' energy because Japan (plans to) reuse its nuclear fuel through the domestic nuclear fuel cycle."[2] The nuclear disaster made it apparent that these discourses were bogus,[3] and many Japanese lost fundamental confidence regarding any claim for the safety of nuclear power.

The reconstruction of disaster-afflicted areas must be undertaken in this completely altered landscape, and this is the context in which reconstruction policies must be devised and implemented. Yet the current policy discussion regarding recovery and reconstruction of the areas severely afflicted by the nuclear accident seems to us frivolous and dubious. For example, at the level of grand design, the Reconstruction Design Council, established by the national government, released on June 25, 2011 a document entitled "Towards Reconstruction: Hope Beyond the Disaster" that listed "seven principles for reconstruction design." Among the seven principles, six focus on recovery and reconstruction from the earthquake and tsunami. Only one refers specifically to the nuclear accident, and it does so only abstractly: "while continuing to seek the near term resolution of the nuclear accident, support and reconstruction of the areas afflicted by the nuclear disaster must be given a more detailed level of consideration" (Reconstruction Design Council 2011, 42). This apparent marginalization of the nuclear accident in the reconstruction discourse is worrisome. At the level of (supposedly more concrete) reconstruction policies for the Great East Japan Earthquake as a whole, the Reconstruction Headquarters in Response to the Great East Japan Earthquake published its Basic Policy for Reconstruction on July 29, 2011.[4] In regard to the nuclear disaster, the seventh point of the section outlining the "basic stance" of this document raises the point that:

> … particularly in regard to reconstruction from the nuclear power plant disaster, the entire country must share a sense of strong crisis, and while this basic policy document sets out urgent actions for recovery and reconstruction, it is not limited to this, but rather takes a long-term perspective on efforts *by the government to continue to take responsibility for revitalization and reconstruction.* [authors' translation; emphasis added by the authors]

It also states that

> … in regard to how to respond to the areas in Fukushima seriously afflicted by the nuclear power plant accident, *while continuing to enforce the conditions of the Atomic Energy Damage Compensation Law and the Act on the Nuclear Damage Liability Facilitation Fund*, the required review process must also be conducted in a manner that corresponds to the conditions of the accident and recovery. [authors' translation; emphasis added by the authors]

A careful reading of this document shows how the government assumes no direct responsibility for the Fukushima Daiichi nuclear accident itself, and only implies its vague responsibility for "revitalization and reconstruction." True, the government is now directly dealing with post-accident issues such as decontamination, interim nuclear-waste storage, and contaminated-water issues, but this is not a sign of taking responsibility for the accident itself. In addition, as implied in the second quote, the government insists on making TEPCO bear direct financial responsibility for the accident, implying its denial of direct responsibility, although it ultimately backs the company by transferring the costs of compensation to ordinary citizens in the form of electricity bills.

Given that the development of nuclear power has been a state-supported enterprise in Japan, we argue that the national government must bear a greater share of responsibility for the accident. Furthermore, if the government is truly committed to the reconstruction of the disaster-afflicted areas and to regained confidence in the country's nuclear policy, a decisive shift away from nuclear-power generation is a prerequisite for reconstruction. We came to this conviction in working with those who are suffering from the disaster and with numerous other local and national actors over the past years. In short, Japan must establish the decommissioning of nuclear power plants nationwide as a policy strategy and clearly outline the timeline for this decommissioning process.

We are clearly aware that this is not the view shared by those currently in power. As a result of the transition of power from the Democratic Party (led by the Kan and Noda Cabinets, who were supportive of nuclear phase-out) to the Liberal Democratic Party (and the nuclear energy-supportive Abe Cabinet), nuclear energy has begun to be repositioned as a vital power source for the nation (cf. Aldrich 2011). Even if this political transition does not lead to the construction of new nuclear power stations, it does suggest that spent nuclear fuel and highly concentrated radioactive waste will continue to be produced. Yet, the prospects for dealing with spent nuclear fuel are daunting. While it has long been planned to undertake nuclear-fuel reprocessing at the Rokkasho Nuclear Fuel Reprocessing Facility in Aomori Prefecture, operation of the facility has been repeatedly delayed. Not-yet-reprocessed spent nuclear fuel remains in a state of intermediate storage at Rokkasho and at nuclear power plants, and surplus storage capacity at these sites is perpetually decreasing. Furthermore, plans for final underground burial facilities for the highly concentrated radioactive waste produced through reprocessing remain completely uncertain, for the simple reason that there are no sites in the highly tectonically active Japanese Archipelago where safe storage for 100,000 years can be guaranteed (Ishibashi 2013). Given this situation, further promotion of nuclear power in the country is nonsensical.

Toward the reconstruction of Fukushima from Fukushima

An institutional framework specifically designed to address the Fukushima nuclear disaster came on March 30, 2012, as the "Act on Special Measures for the Reconstruction and Revitalization of Fukushima."[5] Based on this act, the national Reconstruction Agency released its "Fukushima Reconstruction and Revitalization Basic Guideline" on July 13.[6] What is most notable about this guideline is the role played by Fukushima Prefecture in shaping the document. That is, in the process of drafting this Basic Guideline, Fukushima Prefecture strongly stipulated that its basic principles for reconstruction, outlined in the Fukushima Reconstruction Vision (details below), should be incorporated into the final document. Although local municipalities and prefectural governments have never been completely excluded from nuclear politics in Japan (Aldrich 2013), this can still be seen as an important effort and achievement in exerting control over the process of reconstruction by those who are most affected.

The Fukushima Reconstruction Vision is a set of basic concepts and measures to deal with the still-unfolding disaster that was approved by the prefecture on August 11, 2011.[7] One of the authors (Yamakawa) engaged in discussions of this document as the deputy chair of the Vision commission. The three key concepts are:

1 Building a safe, secure, and sustainable society free from nuclear power.
2 Revitalization that brings together everyone who loves and cares about Fukushima.
3 A homeland we can all be proud of once again.

This prefectural Vision's aim for "a society that does not rely on nuclear power" sent a strong message to the rest of the country as a proposition about the future shape of not only Fukushima Prefecture but of Japanese society. It was also in this Vision that the governor of Fukushima Prefecture requested that TEPCO decommission both Fukushima Daiichi and Daini Nuclear Power Plants (Yamakawa 2012b).

The following principles reflect the Fukushima Reconstruction Vision as well as our collective research efforts and support activities over the past few years (Yamakawa 2012a). Ultimately, these principles are what we believe to be the key ways in which nuclear disaster-afflicted areas should be supported.

Principle 1: rebuilding safety, confidence, and trust

1 "Realization of a nuclear-free society" must become a guiding principle.
2 Decommissioning all nuclear reactors must be made a policy strategy, a clear timetable determined, and new nuclear plants and restart not authorized.
3 Make public all information related to the nuclear accident, damages, predictions, and resolution, and clearly explain that nuclear disasters are human disasters.
4 Creation on a regular basis of detailed radioactive contamination maps in preparation for return, recovery, and reconstruction.

5 Tightening of standards for low-dose external exposure and establishment of a testing framework for dealing with ubiquitous radiation.
6 Distribution of radiation ledgers to residents in the afflicted areas and decontamination personnel, follow-up health checks on long-term effects of low-dose radiation, and full-expenses guaranteed diagnosis and treatment.
7 Establish stigmatization countermeasures for Fukushima products focused on agricultural produce.
8 Presentation of a timetable for the temporary and intermediate storage of radioactive waste stemming from decontamination and restriction of transport of radioactive waste materials to areas within the Futaba District.
9 Prior to complete decommissioning, nuclear-disaster mitigation planning must adopt "escape" as its fundamental principle and all burdens associated with evacuation must be fully guaranteed.

As noted above, one fundamental outcome of this nuclear disaster has been a loss of confidence in the safety of nuclear power and an erosion of trust in the individuals and organizations that promoted the use of nuclear power in Japan: what many refer to as the "nuclear village." This cynical but apt term describes the close ties between the organizations who benefit from nuclear power, including the Ministry of Economy, Trade and Industry (METI), nuclear-reactor manufacturers, electric-power companies, electric-power associations, local governments, and so on (Vivoda 2014). Needless to say, at its root it is the national government that backs these organizations. For example, the National Energy Basic Plan, drafted by METI, is ultimately approved by the Cabinet Office, and the Nuclear Regulation Authority (NRA), even though it is an independent agency, is still a state-supported entity. We should also note that scientists and science have lost their credibility as a result of this disaster (Yamakawa and Yamamoto 2016) because the Science Council of Japan, arguably the most authoritative academy of scientists in the country, ultimately approved the "peaceful use of nuclear power" in the 1950s. To remedy the current situation, information disclosure is absolutely essential (for the lack of it, see Fujimoto 2016). It is equally critical to acknowledge that a nuclear disaster is a human disaster. Beyond this, we argue that the following are necessary means to secure and restore safety, confidence, and trust in the authority.

Our research, which includes a comprehensive survey of evacuees from the afflicted areas, has made it undeniably clear that the foremost concern among disaster victims is radioactive contamination of their living environment (Yamakawa 2016). To respond to this concern, up-to-date, detailed maps of radioactive contamination are essential. These maps must have spatial resolutions fine enough for the residents who decide to stay in or return to their homes to avoid radiation hot-spots in their everyday lives. To that end, various voluntary groups have been playing critical roles, including the radiation-countermeasure team of Fukushima

University, who measured air-radiation doses along streets in the Hama-Dori (coastal) and Naka-Dori (central) regions and made radiation-level maps available to the public.

There are conflicting views on whether or not long-term low-level radiation exposure is harmful, and the scientific evidence is insufficient for the task of proving either view definitively correct. Accordingly, belief in either of these views has turned into a controversy not unlike a battle between religions. For now, and for as long as a conclusion is not reached in regard to the long-term effects of low-level radiation doses on the body, it is most important for residents in afflicted areas to minimize unnecessary internal and external radiation exposure. To that end, close monitoring of the health conditions of residents in the afflicted areas is essential. However, it is not possible to be satisfied with only follow-up surveys, since the results are subject to crafty discursive operations that smooth over anomalies and other serious issues.

For example, in March 2015 the Thyroid Examination Evaluation Division of the Committee for the Fukushima Prefectural Citizens' Health Survey released their Interim Summary of Thyroid Examinations (Committee for the Fukushima Prefectural Citizens' Health Survey 2015). This survey and examination targeted the 300,000 persons under 18 that were living in Fukushima Prefecture at the time of the nuclear accident. The results indicated that 112 persons were found to have thyroid cancers that were malignant or suspected to be malignant. The Thyroid Examination Evaluation Commission confirmed that this frequency was found to be an order of magnitude greater in comparison with statistics for usual incidence of thyroid cancer. However, what needs to be noted is how the text narratives in the interim report are crafted. The report reads: "it can be thought that this frequency is due either to increased radiation exposure or to increased testing (i.e. to discovery of cancer that does not show symptoms or life-threatening prognosis)" (p. 1; authors' translation), thus leaving this higher incidence open to two explanations. The report then goes on to say, "while from the point of view of science the former possibility cannot be negated, the view has been expressed that the potential of the latter is high" (p. 1; authors' translation), without mentioning who expressed this view or the basis on which this particular committee member made this claim. The document later states:

> based on this initial examination, we consider that it is unlikely that the thyroid cancer discovered thus far was caused by radiation since the amount of radiation exposure was far less than in the case of the Chernobyl accident, and because there were no discoveries of thyroid cancer among children under five years old [i.e. who would be more prone to the effects] at the time of the accident.
>
> (p. 2; authors' translation)

This evaluation was determined based on the death-rate results found by the Life Span Study of atomic-bombing victims of Hiroshima and Nagasaki conducted by the Radiation Effects Research Foundation, and it has been criticized by Emeritus

Professor of Nagoya University Shoji Sawada and others for reducing the risks of radiation exposure leading to illness by an order of magnitude (Sawada 2014, 11).

In addition, it is crucial to address the stigmatization (or reputational damage) of food produce that continues to plague Fukushima Prefecture, by developing and implementing strategies aimed at resolving these issues. The key is to establish a system that allows for pervasive and thorough testing of food products for radiation. This is impossible without all of the following steps: creation of detailed maps of the contamination of farmland prior to cultivation; crop selection based on the particular conditions of farmland and crop transfer coefficients during the stage of production; full monitoring at the stage of distribution; and, finally, testing by consumers at the point of consumption (Chapter 6; Koyama and Komatsu 2013).

Dealing with the massive amount of radioactively contaminated soil and waste is another crucial challenge. Local governments' facile acceptance of the construction of interim storage facilities would likely reinforce their structural dependence on "nuclear money," which took the form of a special funding system under the "Three Power-Siting Laws" during the construction phase of nuclear power plants (Samuels 2013; Aldrich 2008).

Nevertheless, the construction of interim storage facilities is already underway in the afflicted areas of Fukushima. Sites for interim storage facilities were selected based on the following five points: 1) ability to secure an adequate area for the site; 2) proximity to areas from which large amounts of contaminated soil and waste will be removed; 3) access to major roads; 4) avoidance of faults or unstable ground; and 5) minimization of changing river flows. A field survey was conducted in April 2013 and in February 2015 in Fukushima Prefecture, along with Okuma Town and Futaba Town, decided to allow interim storage facilities; construction at these sites has been initiated. The storage site in Okuma has a capacity of 10,000 cubic meters. As of September 2, 2015 the actual amount stored was 7,409 bags (each bag holds approximately 1 cubic meter) and the air dose rate was 1–9 μSv/h, thus indicating relatively little change from the period before storage (Ministry of the Environment 2016). However, for evacuees the urgent problem is that final storage-facility sites outside Fukushima Prefecture remain undetermined, and it has thus been impossible to dispel fears that interim storage facilities will become final storage sites.

Principle 2: unburdening victims and evacuees, promoting support that opens prospects for the future

1 TEPCO and the national government must be held accountable for full and non-discriminatory damage compensation and comprehensive life-rebuilding.
2 Full support for funding and training leading to the securing of employment and the resumption of business.

3 Guarantee and expand rights to return from temporary and subsidized housing to former residences and temporary towns, or to relocate to a different site.
4 Implementation of regular health checks and assurance of completely cost-free examinations and treatment for victims.
5 Guarantee the "no cost" and "no burden" provision of middle- and high-school education to all child victims.

There are currently 150,000 evacuees from Fukushima alone living inside and outside the prefecture. Our previous research on evacuees of the nuclear disaster has shown that, besides anxiety over the effects of radioactivity, evacuees are concerned with such issues as financial security, housing, employment, health, and interpersonal relations. Indeed, the focus of victims' and evacuees' concerns is shifting from evacuation to temporary housing, and from radioactivity to finances for daily life, and for future housing and employment. Concerns about how to formulate life plans are particularly prominent for families with children, who are highly concerned with where and how to pursue education. While wondering how long evacuation or life in temporary housing will continue, it is particularly difficult to find any prospects for the future, and such conditions make it difficult for individuals to take steps toward preparing to restart their lives. Moreover, many evacuees have lost their employment and occupations and their living expenses are being covered by donations and provisional compensation payments, receipt of pensions, or devastating drafts on personal savings.

It is the fundamental right of victims who have lost their home and occupation to be adequately compensated by TEPCO for their losses. Thus far, compensation claims to victims in Fukushima Prefecture have been related more to damages from emotional stress ("consolation payments") than to damages to property.[8] Yet the latter compensation is equally, if not even more, critical. In particular, the designation of restricted areas has prevented evacuees from returning home and unrepaired leaks and intrusion by animals have increased structural damages. Moreover, some victims are forced to use their compensation payments simply to continue paying off a loan for a house in which they cannot live. Asset inequality is exposed in the ability to restart life. This asset inequality will eventually lead to educational differences in successive generations, and the wealth gap will be dragged into the future as social DNA. For these reasons, compensation payments are absolutely essential to evacuation and rebuilding.

Other support, not in the form of direct payment, is also essential, including employment training, funding for business restart, flexible support for evacuees regardless of their eventual destinations, cost-free health support for victims, and education opportunities for child victims. In the provision of these compensations there are several major considerations that require caution—a fact that became apparent based on our research and support experience over the past years, as described below.

First, we must recognize that some compensation payments have also become a source of division among victims. For example, in June 2011, sites in Date City and Minami-Soma City where cumulative external radiation exposure was found to be over 20 mSv/year were declared "specific evacuation encouragement zones." The boundaries of these evacuation zones pierced through communities (although, as already mentioned above, there may be a number of radiation "hot-spots" if measured with more precision). The result is a seemingly unreasonable gap in compensation payments between neighboring households, and emotional conflicts among them. Attention to and care for such potential divisions must be taken into account when establishing compensation schemes.

Our second point is associated with the discrete and dispersed nature of evacuation in the nuclear disaster (see Yamakawa 2016). Due to the unusually large extent of damages from the Great East Japan Earthquake Disaster, which resulted in a shortage of temporary housing, and because of discrete forms of evacuation that followed the nuclear accident (e.g. family members evacuating to and living in different locations), a new form of evacuation housing was institutionalized. The so-called *minashi* ("deemed") temporary housing essentially consists of *government-subsidized rental units*, which are normally commercial apartments, but have come to be designated as evacuation housing and subsidized by local governments. This "deemed temporary housing" played an important role following the earthquake disaster.[9] However, the emergent side effect of this "successful" housing is that it has become rather difficult to check on and obtain information from the evacuees living in these housing units scattered across the area, especially if they are outside the prefecture.

Third, there are critical suggestions that many current government policies are "forcing" (or strongly incentivizing) evacuees to return to their homes, and are unsupportive of residents who will not return. While national, prefectural, and local governments have said little more than that it is "difficult to return," there are numerous cases of evacuees who have given up on their former houses in the "difficult-to-return areas" in favor of starting a new life, after seeing the desolate state of their home and its surroundings during their temporary visits. We must not presume, therefore, that rebuilding afflicted areas by bringing former residents back is always the correct stance. Rather, government agencies and officials must first try to understand why some residents return and some do not, respect their decisions, and provide fair compensation and institutional support regardless of the evacuees' ultimate decisions on residence.

Last, we must recognize that compensation payments also affect evacuees' decisions to return to their former communities. In Hirono Town and Kawauchi Village, for example, even after the "emergency evacuation preparation area" designation was lifted, and indeed even a full year after government offices were returned to the towns, the return of residents has not proceeded apace. Why? Surely there are problems with the still unfinished project of decontamination, delays in the recovery of water and sewerage services, difficulties in finding employment, and insufficiencies in medical services, schools, and retail shops. However, the issue of compensation is also a factor. We have heard individuals privately voicing the opinion that as long as subsidized rentals are available, they would rather not hurry to return.

Principle 3: revitalization of local identity

1 Focus on the restoration of the original landscape, cultivated through nature–human interactions over time, rather than a "creatively reconstructed" landscape.
2 Preservation of unique local traditional and cultural values and the continuation of festivals.
3 Community revitalization centered on establishing local community associations (temporary housing residents) and wider regional associations (government-subsidized housing residents and evacuees living outside Fukushima Prefecture).

The local identity of an inter-subjective community is a co-production of humans and nature formed over time through the cultivation of natural, architectural, and cultural environments that integrally reinforce the framework of the local (Vidal de la Blache 1922; Ueno 1972). Since what many residents of the evacuated areas are hoping for is the restoration and revitalization of the home communities that were once the taken-for-granted but essential basis of their local identity, they are opposed to "creative reconstruction" plans that completely transform the locality itself (Yamakawa 2016). Indeed, proponents of "creative reconstruction" often raise the need to conduct these operations in a timely manner (also see Chapter 4). However, it is known that if diverse opinions are not respected and proper deliberations not ensured, the pushing through of top-down style leadership can lead to unnecessary disruptions in consensus-formation. Even if it takes time, deliberation is essential. If conclusions can be reached after adequate deliberation, then the concretization of recovery and reconstruction procedures can be advanced in a truly "timely" manner. This is the fundamental perspective that is proposed in the second fundamental concept of the Fukushima Reconstruction Vision, which reads: "revitalization that brings together everyone who loves and cares about Fukushima." It is here that humanities and social-science scholars can be expected to make inroads into efforts to support disaster reconstruction.

The third pillar of the fundamental concept of the Fukushima Reconstruction Vision is "revitalization that realizes communities overflowing with pride." In life, pride in being an individual human person is critical and, to reinforce and strengthen this pride, it is essential to rebuild local identity. In the effort to rebuild local identity, traditional culture is essential. Whether the specific form that this traditional culture takes is late-summer *bon-odori* (community dancing), the *shishi-odori* (lion dance), or the *nomaoi* (wild horse-chase festival), it is impossible to ignore the restorative power of traditional culture. When we consider the fact that these traditions and cultures have been handed down through the past, it becomes easy to understand why they are so important to the future.

The rebuilding of community ties is also essential. Here we would like to consider how to reconnect and restore these ties. As an example, the "mutual help center" in Tomioka Town has developed such activities as a human-resource center, crafts, farming, and a restaurant in order to "rebuild a sense of fulfillment and the living environment." Furthermore, residents living in government-subsidized rental housing have been supported through the utilization of tablet terminals and other electronic devices. It is necessary to restore the former ties between all of the people and communities who have been separated by this disaster. It is particularly important to find means of recuperation for the many individuals who evacuated outside the prefecture and were subsequently labeled "deserters" by those who stayed behind. Such labels and feelings turn into a great psychological weight that can prevent many residents from ever returning. In order to lower this psychological wall it is imperative to open lines of communication between those who left and those who stayed, such that they can begin to understand that they both have faced the same concerns. This is a necessary and important opportunity for people who left to reduce their psychological burdens. Ill-feeling between children can also be dispelled through the opportunity to play together.

Principle 4: revitalization-oriented community-based planning

1 Strengthen and expand the capacity of local government offices by increasing the number of officers and strengthening collaboration between local governments through, for example, the "pairing system."
2 Facilitate space for promoting collaborative community-building efforts that are a product of the citizens, industry, academia, and government.
3 Establish focal points for exchange through health, retail, sports, culture, and learning based on the autonomy and spontaneity of local residents.
4 Establish one-stop and on-demand systems for all financial, postal, home-delivery, and shopping services, as well as medical, nursing, and welfare facilities.
5 Promote public reconstruction housing policies tailored to the reality of nuclear disaster and evacuation, and secure the right of the evacuees to participate in community development in both their original hometowns and evacuation destination communities.

In the Reconstruction Design Council's plans for reconstruction measures, local governments are expected to play a pivotal role in simultaneously pursuing post-disaster reconstruction and *machi-zukuri* ("town-making" or community development/planning; see Sorensen and Funck 2007) by coordinating "mutual support" efforts to fill in gaps that cannot be addressed either by "self-help" or "public assistance." In the Council's words:

> Local governments must present the options for reconstruction, including the various pros and cons of each option, to their citizens, and make decisions about directions to be taken based on listening to a wide range of opinions from local residents and stakeholders.
>
> (Reconstruction Design Council 2011, 10)

However, many of the afflicted rural communities were suffering from the erosion of their governance capacity even before the disaster, due to such factors as aging, depopulation, and low birth rates. What has become plainly evident through the Great East Japan Earthquake Disaster is that the weakening of local government and administrative capacities has severely delayed evacuation, recovery, and reconstruction efforts (also see Chapter 5 for related discussion in the context of transportation planning). Therefore, disaster response and community development in the afflicted areas must start by rebuilding the capacities of local governments. One possible way of achieving this is the establishment of a "pairing system" whereby municipalities make agreements to provide various forms of mutual assistance, including sending personnel and other resources, in the case of disaster (Samuels 2013). In fact, these pairing systems were already in place among many municipal governments during the disaster, but some did not function well, and improvement of these systems is necessary (Sato and Miyazaki 2015).

In addition to strengthening local governments' administrative capacities, it is critical to complement these capacities with volunteer and non-profit organizations, private industries, and academia. There are already numerous citizen-led *machizukuri* movements in Japan (Sorensen and Funck 2007). Most recently the idea of "Machizukuri Companies" has been proposed as a new form of non-profit organization, modeled after Business Improvement Districts in the United States, that would work in partnership with public institutions. This disaster also reinforced the fact that private industries may play a critical role in reconstruction. For example, a national convenience-store chain established a store in Kawauchi Village as soon as reconstruction efforts began there (Chapter 9). These stores not only sell food and daily goods, but also provide various other services such as ATM machines, bus-ticket sales, and utility-bill payments. Most of the individuals who have returned to Kawauchi are elderly, and the store provides vital services for them. The Great East Japan Earthquake Disaster also demanded a major change in the roles played by academia in the reconstruction of afflicted areas. This is a shift from "research for the sake of research" to "research for society." It is also a shift from social experiments to social practice. Many post-secondary academic institutions established not disaster-research centers but volunteer-activity centers and reconstruction-support centers, as seen in the establishment of FURE at Fukushima University.

In strengthening the administrative capacity of local offices, the most important aim must be to increase capacity for communication with evacuees. The ability to do this will greatly depend on whether or not it will be possible to take into account the diverse opinions of all victims and evacuees. Local-government plans for reconstruction are decided on the basis of local questionnaire surveys, but these surveys are prone to bias toward the views of a particular subset of victims. If only the head of the household is targeted by these surveys, then the results will be strongly biased toward the opinions and perspectives of middle-aged males and it will remain difficult to incorporate the opinions of women and the younger generation into reconstruction planning. Also, it has been pointed out that it is quite difficult to gather the "true" feelings of respondents through questionnaire surveys. As evidenced in the case of Tomioka Town, it is possible to overcome these obstacles through regular and frequent town meetings (Mori, Shirafuji, and Aikyo 2012).

Evacuees are forced to choose whether to return to their former communities or to build new communities away from home. The problem with returning is that former communities have been stripped bare of life, services, and infrastructure as a result of radioactivity. Evacuees are comparing and contrasting life at their evacuation destinations with the desolation of their former communities, which often leads them to reevaluate returning. These comparisons are not only about electricity, gas, sewers, gasoline, food, and other basic infrastructure, but also about medical, health, education, retail and cultural services, as well as employment and other necessities for a decent life. Furthermore, we must recognize that these services not only function to facilitate individual well-being but also facilitate social interactions among the residents, an essential aspect of quality of life. Restoration efforts should not, therefore, overlook the importance of establishing focal points for these welfare-enhancing and socially critical activities.

Unlike most disasters experienced by Japan in the past, in this case many medical personnel left the afflicted areas due to fears of radiation exposure, and truck drivers were barred by their employers (or voluntarily abstained) from entering evacuation zones. Accordingly, individual or community efforts alone were simply insufficient to secure the necessary services and goods to sustain their lives and livelihoods. For that reason, the establishment of one-stop and on-demand systems for all financial, postal, delivery, shopping, and medical services is essential, ideally at the scale of the elementary-school district.

Originally the residents of evacuation-designated areas (limited-residence areas and difficult-to-return areas) were supposed to live in temporary housing for up to two years, based on the Disaster Relief Act. However, even five years after the nuclear accident, some of these residential restrictions have not yet been lifted, and many residents are expecting to live in temporary housing for some time to come. The temporary housing is not only small but also is now suffering from structural deterioration, creating physical and mental distress for residents. To amend the current situation, policymakers must first move away from the orthodox method of building simple prefabricated houses and instead build more permanent, public reconstruction housing from the beginning. Second, because public reconstruction housing requires relatively large plots of land, local governments should secure large amounts of public land as part of their disaster-prevention and response-planning policies. Third, the institution of policies that facilitate the building of smaller public reconstruction housing units in dispersed locations is urgently needed (Yamakawa 2016). This form of dispersed small reconstruction housing was found to have multiple benefits in this disaster, including convenience for residents because it can be built on small plots of urban areas with good access to transportation and services and has the potential to mitigate the problem of "hollowing out" of central commercial districts, and, if built in rural areas, the potential to maintain existing communities by keeping residents close to their original homes.

Finally, in addition to these infrastructural issues, it is critical to encourage evacuees to participate in community-development efforts and local policymaking both in their evacuated and their evacuation communities. Currently the Japanese government is demanding that evacuees choose either to "return home" or "relocate (permanently)" through its Early Return and Settlement Plans. The Science Council of Japan has instead insisted that evacuees should be allowed the option to "wait and see" (i.e. not make a decision now and continue evacuation), and has proposed the institutionalization of the "dual residence registry" system (Science Council of Japan 2014).

Principle 5: decommissioning and renewable energy as basis for sustainable transition

1 Adoption of "scenario zero" as energy strategy; stoppage of nuclear restart and establishment of new nuclear power plants.
2 Work toward the establishment of technologies for decommissioning nuclear reactors; strengthen collaboration with international research agencies and establish sites for training individuals who will be responsible for such activities.
3 Create sites for the manufacture and assembly of equipment for renewable energy while also training personnel in how to maintain such equipment, and develop facilities for education and training towards such ends.
4 Enhance the power grid through the separation of electrical-power production and power distribution and transmission, and use the purchase obligation and fixed-price system of the nine power companies to promote local production and consumption of energy.
5 Reconsider local resources such as agriculture, forestry, and fisheries, and transition to land uses that will increase national self-sufficiency for food.

Prior to the TEPCO nuclear disaster, national energy policy focused on "the pursuit of economic efficiency (lower costs)," "energy security (quasi-domestication)," and "adaptation to the environment (reduction of CO_2)." In line with these perspectives, efforts to "green" industry focused on reducing carbon-dioxide emissions by increasing the nation's reliance on nuclear energy to 50 per cent by 2030. However, after the TEPCO nuclear disaster, the perspectives of "safety and reliability" were added to the national energy strategy. Subsequently, three energy supply-structure scenarios were created for 2030 and public debate initiated. These three energy scenarios set the reliance on nuclear power at 0 per cent, 15 per cent, and 20–25 per cent respectively. The following general perspectives are used to outline an image of energy policies in 2030: "securing the safety of nuclear energy," "strengthening of energy security," "contributing to resolving global warming," and "limiting costs and preventing hollowing out." Our position is clearly to support the first, 0 per cent nuclear scenario ("scenario zero").

Multiple reactors falling into a crisis state is a world historical first.[10] Accordingly, the international nuclear agencies such as the International Atomic Energy Agency (IAEA), as well as the nuclear industry, continue to be indirectly involved in advising on such issues at the nuclear-accident site as the removal of debris, demolition of the reactor units, and the disposal of both; they also continue to carefully watch the efforts of the Japanese government and TEPCO, as well as decommissioning operations. Since decommissioning operations can only be carried out at the accident site, related agencies and operators will need to establish cutting-edge international research-and-development facilities in the vicinity of Fukushima Daiichi. However, even if such facilities are brought to the area, actually putting the fruits of their research and development into practice at the accident site will require training human resources capable of carrying out these operations. It will also be necessary to secure the safety of these workers. Such education and training facilities for human resources must be established.

One new industry that is expected to play a vital role in reconstruction is renewable energy (also see Chapter 10). Renewable energy can be expected to serve as a replacement for nuclear energy, and since it is a smaller-scale and more dispersed industry, it is amenable to local revitalization and essential to reconstruction and recovery. It is also necessary to ensure safety and security through the pursuit and integration of multiple forms of renewable energy—solar, wind, small-scale hydropower, and biomass—in order for the local production and consumption of energy to provide a strong base for energy policy as a whole. To select an energy strategy is in many ways to select the mode of the local life and economy, and also to select the mode of humanity (Okada 2012). If nuclear phase-out becomes a clear policy stance, as in the case of Japanese automobile manufacturers who were able to adjust to the regulation of exhaust gases through the Clean Air Act, then industry can also begin to develop and adjust in line with decommissioning and greening. As part of efforts to curb global warming, renewable-energy generation and transmission, and technologies such as "smart meters" aimed at decreasing energy consumption, have already been promoted, and should be pushed forward further.

Japan's spatial economic structure has become increasingly centralized around Tokyo since the 1980s. During past major disasters such as the Hanshin Awaji Great Earthquake (1995) and the Chuetsu Earthquake (2004), and in the most recent Great East Japan Earthquake Disaster, the vulnerability of manufacturing supply chains, such as those of the auto industry, in which production points are highly specialized and concentrated has become apparent. Similarly excessive reliance on Fukushima Daiichi and Daini Nuclear Power Plants for Tokyo's electricity resulted in a shortage of electricity in the Tokyo metropolitan area during the summer of 2011. We are already observing emergent corporate strategies to disperse their production locations, such as Toyota's plan to build another production complex in Miyagi in addition to the two existing ones in Aichi and Fukuoka. In the realm of energy provision, we must establish a system of policies encouraging public management of the power grid and the involvement of various forms of power provision and sales in order to develop a dispersed system of local production and local consumption of energy (Chapter 10). This transition to renewable energy holds the potential to strengthen beneficial feedback between various human practices and nature in local economies, and will become the driving force behind the promotion of localization of the production and consumption of water, food, and other resources.

Many of the same points can be made in regard to food supply. The areas affected by the Great East Japan Earthquake Disaster are dependent to a considerable degree on primary industries. In the coastal areas, trends in fishing and seafood processing as well as tourism greatly shape the state of the local economy, while in agricultural areas rice and vegetable production greatly impacts the local economy. Accordingly, reconstruction and recovery from the triple earthquake, tsunami, and nuclear disasters demands the revitalization of primary industries (also see Chapters 6 and 7). Since the recovery of primary industries contributes not only to increased GDP but also to the quality of life in the region, the revitalization of primary industries is an essential source for the cultivation of a spirit of revival in an aging local society.

Conclusion

Inhabitance is fundamental to human life. If inhabitance cannot be restored or is reduced to the level of mere biological survival, this violates the Constitution of Japan's guarantees of the "right to the pursuit of happiness" (Article 13) and the "right of abode" (Article 22). In this chapter, we consider what actions and measures must to be taken to secure nuclear-disaster victims' right to inhabitance. The right to happiness guaranteed by the Constitution of Japan is not the right to happiness in some outstanding sense. It is, rather, the right to a normal everyday life right now. Evacuees' desire to have their pre-disaster lives restored is nothing more than the exercise of this right. To that end, proper compensation for the victims from the government and TEPCO is absolutely essential, as is tailoring post-disaster reconstruction policies to the distinct nature of the nuclear disaster and forms of evacuation. We cannot emphasize enough, nevertheless, that the most important stance underlying our propositions of the principles of reconstruction is to respect, ensure, and cultivate the restorative will and strength of the afflicted people and communities.

Our vision for the future, including the shift toward increasing renewable energy, should be seen as an extension of this basic stance. We are staunchly against a top-down "creative reconstruction" approach as advocated by some public officials and business elites, but this does not mean that we favor the status quo, either. Indeed, our vision will require the introduction of novel spatial planning, industrial, regional, and social policies that can facilitate the following shifts: the development of communications and information technology to enable a shift from an industrial structure of large consumption toward one premised on low energy; from a type of industrial location centered on an over-concentrated pyramid structure to a multi-pole, flatter structure; from community-building reliant on private automobiles and sprawl to compact communities based on public transportation and walkability; from over-concentrated systems of energy production and provision to the unified management of local production and consumption at multiple sites; and from a mode of daily life predicated on mass consumption to low energy. Finally, it is necessary to facilitate population movements from urban to rural areas. We understand that none of these goals will be realized immediately, but firmly believe that such ideals are essential guides toward a more just and sustainable society—one that will overcome the range of daunting challenges brought upon us by the historic disaster.

Notes

1 Center homepage: http://fure.net.fukushima-u.ac.jp/ (accessed April 19, 2016).

2 This term refers to the process of extracting plutonium through reprocessing spent nuclear fuel after nuclear reaction and "re-using" it for power-generation. Japan has to import the raw materials used in the production of the uranium fuel rods essential to generating nuclear power, but proponents of this argument suggest that because the extracted plutonium is not imported, nuclear power may be interpreted as (almost) domestic energy.

3 For the first discourse, a severe accident did happen. For the second discourse, if a nuclear accident occurs, even if nuclear-energy use was peaceful, radioactive contamination results in the same way as with nuclear weapons, such that it is impossible to distinguish between war and peace (Sano 2014). For the third discourse, since uranium must be imported, we are certainly not talking here about "domestic energy" in the original sense of the term. Moreover, the fact is that the fast-breeder reactor at the Monju Nuclear Power Plant was troubled by accidents and is already scheduled for decommissioning, thus demonstrating that the fantasy of the plutonium cycle is a long way from reality. While plutonium was born of the notion of "peaceful use," it can at any time be converted into nuclear weapons for military use and is not only at high risk of becoming a target of international terrorism but also brings fear regarding whether Japan is actually arming itself with nuclear weapons.

4 The official title of the policy is the "Basic Policy for Reconstruction in Response to the Great East Japan Earthquake." This policy was based on the Basic Act on Reconstruction in Response to the Great East Japan Earthquake (June 24, 2011).

5 Act on Special Measures for the Reconstruction and Revitalization of Fukushima (Act No. 25 of March 31, 2012). See www.japaneselawtranslation.go.jp/law/detail/?id=2582&vm=04&re=01&new=1 (accessed September 13, 2016).

6 Fukushima Reconstruction and Revitalization Basic Guideline (July 13, 2012). See www.reconstruction.go.jp/topics/001084.html (accessed April 19, 2016).

7 See www.pref.fukushima.lg.jp/download/1/vision_for_revitalization.pdf (accessed April 19, 2016).

8 Consolation payments of 100,000 yen (about US$ 830) have been paid monthly to individuals forcibly evacuated from the "restricted areas," and monthly consolation payments of 80,000 (US$ 670) yen have been paid to individuals outside the restricted areas but in areas with radiation levels over a certain limit (however, pregnant women and children have received 400,000 yen (US$ 330) per month).

9 The enormous number of victims of the earthquake, tsunami, and nuclear disasters created a housing situation that simply could not be resolved through the construction of emergency temporary housing. In retrospect, the decision made in the early days after the initial nuclear accident to designate housing that met certain conditions as government-subsidized private rental units, or "deemed temporary housing," was extremely wise. In any future cases of a wide-area disaster with large numbers of victims and evacuees, the provision of "deemed temporary housing" may again prove to be a useful technique.

10 The nuclear accident at TEPCO Fukushima Daiichi involved four nuclear reactors, units 1–4. Currently, the removal of nuclear fuel has only been completed at reactor unit 4. Reactor units 1–3 all remain in a state of meltdown. While cold-shutdown conditions are in place at these reactors, dissolved nuclear-fuel debris remains at the bottom of the containment vessels. As these conditions clearly evidence, the results of the accident have not been completely resolved. Moreover, the complete resolution of the accident's results is not on any foreseeable horizon.

References

Aldrich, Daniel P. 2008. *Site Fights: Divisive Facilities and Civil Society in Japan and the West*. Ithaca, NY: Cornell University Press.

Aldrich, Daniel P. 2011. "Future Fission: Why Japan Won't Abandon Nuclear Power." *Global Asia* 6(2): 62–67.

Aldrich, Daniel P. 2013. "A Normal Accident or a Sea-Change? Nuclear Host Communities Respond to the 3/11 Disaster." *Japanese Journal of Political Science* 14(02): 261–276.

Committee for the Fukushima Prefectural Citizens' Health Survey. 2015. *Interim Summary of Thyroid Examinations*. Accessed March 10, 2016. www.pref.fukushima.lg.jp/uploaded/attachment/115335.pdf. [In Japanese]

Fujimoto, Noritsugu. 2016. "Decontamination-Intensive Reconstruction Policy in Fukushima Under Governmental Budget Constraint." In *Unravelling the Fukushima Disaster*, edited by Mitsuo Yamakawa and Daisaku Yamamoto, pp. 106–119. London: Routledge.

Ishibashi, Katsuhiko. 2013. "Can Proper Site of HLW Geological Disposal Be Selected in the Tectonically Active Japanese Islands?" *Gakujutsu no Doko* [Trends in the Sciences] 18(6): 27–33. [In Japanese]

Koyama, Ryota, and Tomomi Komatsu, eds. 2013. *Nou no Saisei to Shoku no Anzen: Genpatsu Jiko to Fukushima no Ninen* [Recovery of Agriculture and Safety of Food: Two Years of the Nuclear Accident and Fukushima]. Tokyo: Shin Nihon Shuppansha. [In Japanese]

Ministry of Environment. 2016. *Chukan Chozou Shisetsu no Gaiyou* [The overview of the interim storage facilities]. Accessed January 16, 2016. http://josen.env.go.jp/chukan-chozou/about/. [In Japanese]

Mori, Hideki, Hiroyuki Shirafuji, and Koji Aikyo. 2012. *3.11 to Kenpo* [3.11 and the Constitution]. Tokyo: Nihon Hyoron Sha. [In Japanese]

Okada, Tomohiro. 2012. *Shinsai Kara no Chiiki Saisei: Ningen no Fukko ka Sanji Binjogata "Kozo Kaikaku" ka* [Regional Rebuilding from the Disaster: Human Redevelopment or "Structural Reforms": Capitalizing on Disasters]. Tokyo: Shin Nihon Shuppansha. [In Japanese]

Reconstruction Design Council. 2011. *Fukko eno Teigen: Hisan no naka no Kibo* [Towards Reconstruction: Hope Beyond the Disaster] (June 25, 2011). Accessed May 20, 2016. www.cas.go.jp/jp/fukkou/pdf/kousou12/teigen.pdf [in Japanese] and www.cas.go.jp/jp/fukkou/english/pdf/report20110625.pdf. [Provisional English translation]

Samuels, Richard J. 2013. *3.11*. Ithaca, NY: Cornell University Press.

Sano, Masahiro. 2014. "Historical Shaping of Social Image of Atomic Power." *Gakujutsu no Doko* [Trends in the Sciences] 19(3): 56–63. [In Japanese]

Sato, Motohiro and Tsuyoshi Miyazaki. 2015. "Saigai to Jichitai no Kyouryoku Kankei" [Disaster and Cooperative Relationships between Local Governments]. In *Shinsai to Ekonomii* [Disaster and Economy], edited by Makoto Saito, pp. 3217–3246. Tokyo: Toyo Keizai Shimpo Sha. [In Japanese]

Sawada, Shoji, Michiyuki Matsuzaki, Katsuma Yagasaki, Susumu Shimazono, Kosaku Yamada, Hyoji Namai, Kanna Mitsuta, Obuko Koshiba, and Masato Tashiro. 2014. *Fukushima eno Kikan wo Susumeru Nihon Seifu no Yottsu no Ayamari* [Four Errors by the Japanese Government that Promotes the Return to Fukushima]. Tokyo: Junposha. [In Japanese]

Science Council of Japan. 2014. *Tokyo Denryoku Fukushima Daiichi Genshiryoku Hatsuden Jiko ni yoru Chouki Hinansha no Kurashi to Sumai no Saiken ni Kansuru Teigen* [Recommendations on the Reconstruction of Livelihood and Housing of Long-Term Evacuees of the TEPCO Fukushima Daiichi Nuclear Power Plant Accident] (September 30, 2014). Accessed March 14, 2016. www.scj.go.jp/ja/info/kohyo/pdf/kohyo-22-t140930-1.pdf. [In Japanese]

Sorensen, Andre, and Carol Funck (eds). 2007. *Living Cities in Japan*. London: Routledge.

Ueno, Noboru. 1972. *Chishigaku no Genten* [The Origin of the Regional Geography]. Tokyo: Taimeido. [In Japanese]

Vidal de la Blache, Paul. 1922. *Principes de Géographie Humaine, publiés d'après les manuscrits de l'Auteur par Emmanuel de Martonne*. Paris: Armand Colin.

Vivoda, Vlado. 2014. *Energy Security in Japan: Challenges after Fukushima*. Farnham: Ashgate Publishing.

Yamakawa, Mitsuo. 2012a. "Thoughts on the Grand Design for Reconstruction of the Areas Affected by the Nuclear Disaster." In *Thinking about Fukushima Now: Issues Related to the Earthquake and Nuclear Disaster and the Responsibility of Social Science*, edited by Y. Goto, K. Morioka, and K. Yagi, pp. 133–166. Sakurai: Shoten. [In Japanese]

Yamakawa, Mitsuo. 2012b. "Toward the Fukushima of No Nuclear Power Station; Reason for Hanging Up Nuclear Power Phase-out over a Revival Vision." *Sekai* 829: 119–129. [In Japanese]

Yamakawa, Mitsuo. 2016. "Living in Suspension: Conditions and Prospects of Evacuees from the Eight Municipalities of the Futaba District." In *Unravelling the Fukushima Disaster*, edited by Mitsuo Yamakawa and Daisaku Yamamoto, pp. 51–65. London: Routledge.

Yamakawa, Mitsuo, and Daisaku Yamamoto (eds). 2016. *Unravelling the Fukushima Disaster*. London: Routledge.

2 International efforts to support disaster risk reduction

Satoru Mimura

The lessons of the Great East Japan Earthquake Disaster

Japan first received support from abroad after a major disaster following the Kanto Earthquake of 1923, when more than 50 countries provided support (Ministry of Foreign Affairs of Japan 2014). More recently, after the Hanshin Earthquake in 1995, 44 countries and regions provided support for search-and-rescue operations, as well as medicine and other supplies (Nishikawa 1996). In the aftermath of this disaster, the government took decisive action toward receiving aid from abroad. However, due to limited past experience with receiving foreign disaster-relief support, institutional frameworks for facilitating the reception of foreign aid were found to be lacking and the government ended up declining some offers of assistance. As such, the need to develop frameworks for facilitating international disaster-relief efforts was recognized as an issue needing to be addressed.

In the case of the Great East Japan Earthquake Disaster, human-resource support in the form of search-and-rescue and medical personnel was received from 24 countries and regions. Additionally, 126 countries, regions, and organizations provided supplies and donations. While Operation Tomodachi, a large-scale relief effort conducted by over 20,000 US military forces stationed in Japan, received the most attention, supplies and donations were also received from developing countries in Asia and Africa (Cabinet Office of Japan 2011). It should also be noted that support from abroad came not only in the form of human and financial assistance, but also in a psychological register. The fact that people around the world made gestures of encouragement and support for the Tohoku region played an important role in caring for the mental states of the victims (Ranghieri *et al*. 2014).

In comparison with the Hanshin Earthquake Disaster of 1995, the reception of assistance from abroad went more smoothly following the Great East Japan Earthquake Disaster. As a result of the accumulation of experience with large-scale disasters at home and abroad during the 16-year period separating these disasters, Japanese and foreign rescue teams were able to coordinate in the field, thus indicating great progress in the right direction. However, with regard to matching offers for help from abroad with local governments in Japan, and particularly with regard to how to adapt to the continually changing needs that evolve in the months and years after a disaster, further improvement in the effective reception and utilization of foreign aid is needed (Nishikawa 1996; Katayama 2013).

As the recent experiences of Japan and many other countries indicate, international cooperation in post-disaster support has become increasingly common. However, even for Japan, a country with ample experience in disaster risk reduction (DRR), there remain many challenges and difficulties, particularly with regard to the numerous challenges imposed by the Great East Japan Earthquake and the Fukushima nuclear accident. In this chapter, I examine why international cooperation in DRR is increasing in importance, how international frameworks for DRR evolved in recent decades, and prospects for international cooperation in pre-disaster investment in DRR, with a particular focus on the Build Back Better approach.

Disaster as a global issue

There are at least three reasons for the growing recognition that international cooperation is essential in facilitating pre-disaster support and post-disaster recovery efforts. First is the increasing number and magnitude of disasters worldwide. As indicated in Figure 2.1, the number of disasters occurring around the world, and their associated damages, have dramatically increased in recent years. As this evidences, the degree of damage resulting from a disaster is not determined by the scale of a natural phenomenon or the size of an accident alone, but is also greatly influenced by the resilience and vulnerability of the society that is presented with these hazardous conditions. For example, the 2010 Haiti earthquake devastated buildings in Port-au-Prince including landmarks such as the Presidential Palace and the National Assembly building, while in Sendai in 2011, where seismic

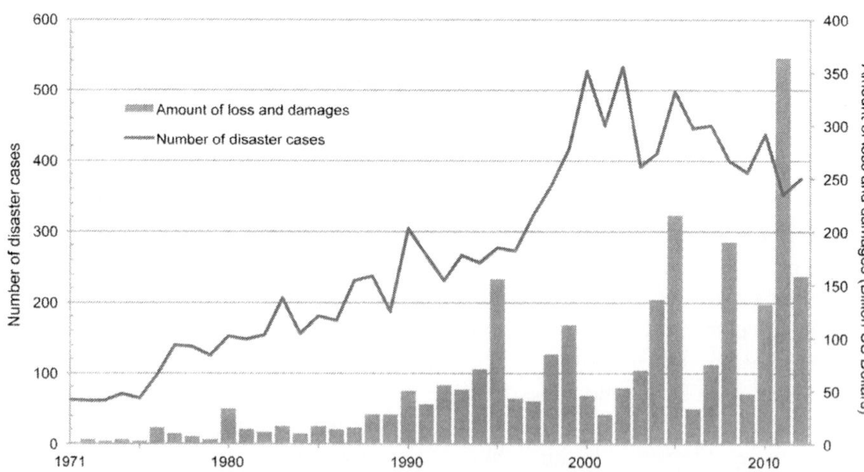

Figure 2.1 Number of disasters in the world and associated damages.

Source: Compiled from EM-DAT.

intensity was larger than in Haiti, no major quake-related damage was reported. This was due to the presence of highly seismic-resistant construction techniques and regulations in Japan (Fujimoto 2014). In short, the scale of a disaster is determined in the interplay between the "external force" of a natural phenomenon or accident and the internal capacities for "resistance and response" of a particular society (Kawada 2003).

Between 2000 and 2011, natural disasters resulted in 1.1 million deaths, 2.7 billion victims, and damages totaling 1.3 trillion dollars (United Nations International Strategy for Disaster Reduction (UNISDR) 2012). Behind the increasing prevalence of powerful and damaging storms can be found a number of important shifts, including climate change resulting in more frequent and larger typhoons, changes in rainfall patterns and other global climatic changes, and increased social vulnerability stemming from population growth and urban over-concentration. The global population has doubled since 1950, and humanity is now extending its habitations into floodplains and steeply sloped areas that are highly vulnerable to disasters (Yanagisawa 2013). Moreover, most people living in these vulnerable areas are poor (UNU-EHS 2014). This phenomenon of poorer classes living in vulnerable areas is particularly prominent in developing countries experiencing rapid population increases but, as demonstrated by the flooding in New Orleans in 2005, even in developed countries it is the poorer classes that bear the brunt of a disaster.

Second, due to the globalization of economies over the course of recent decades, a disaster in one country can lead to substantial economic paralysis, with major impacts on global production. For example, in 2011, the Great East Japan Earthquake Disaster and the Thailand floods heavily impacted global supply chains in the area of auto production, halting production in countries around the world and producing turbulence in the global market (UNISDR 2013). Herod (2011) describes what happened soon after the earthquake in Tohoku:

> By 16 March 2011, all twelve automakers in Japan had temporarily stopped production at some of their plants in an effort to conserve electricity. Within 1 week of the earthquake Toyota's shutdowns had cost it 140,000 vehicles, while Honda lost output of an estimated 47,000 cars and 5,000 motorcycles. Japanese producers also began slowdowns in their US plants due to concerns over the availability of parts shipped from Japan.
>
> (ibid., 831–832)

Herod (2011) furthermore shows that the impact of such disasters as the Great East Japan Earthquake is not limited to the production and exports of commodities; rather, it also shows "how the planet is connected in terms of financial flows" (ibid., 833). Indeed, at the outset of the disaster,

> … anticipating that Japanese firms and the government would curtail or even repatriate overseas investments so as to have the capital available to finance the rebuilding of devastated areas, as well as to pay out insurance claims,

> international investors pushed the value of the yen to a record ¥76.25 per US dollar on 17 March. Only after monetary authorities in Japan, the USA, the EU, Canada, and Britain agreed to intervene in currency markets did its value return to what it had been before the tsunami hit.
>
> (ibid., 833)

Accordingly, while internationally based disaster-recovery efforts have until recently been conducted primarily from a humanitarian perspective, the importance of these efforts from an economic perspective is also increasing.

Third, given the increased movement of people around the world, there are many cases in which victims of a disaster are foreign nationals. For example, in the earthquake disaster that struck Christchurch, New Zealand in February of 2011, many of the individuals killed or injured by the collapse of buildings were foreign students from Japan, China, and other countries (New Zealand Police 2012). While the scale of the disaster was of a level that a developed country like New Zealand could potentially handle on its own, the government accepted support from Australia, Japan, China, and other countries as many foreigners were killed or affected, and announced to the world that this earthquake was not only a disaster for New Zealand.

These factors contribute to the increasing awareness that large-scale disasters are no longer confined to the internal affairs of a single country, but rather present issues that must be approached through international cooperation. Accordingly, frameworks and rules to support these international disaster-recovery efforts are essential.

International frameworks for disaster reduction

Since the late 1980s there have been various efforts to establish frameworks and rules in DRR through a number of international conferences. Below I examine the evolution of these international collaborations over policies and action frameworks for DRR by focusing on key international events and established organizations, including the Hyogo Framework for Action in 2005 and the Third World Conference on Disaster Risk Reduction in 2015. Of particular interest here is the change in underlying assumptions in international DRR during the past decade.

At the 42nd session of the General Assembly of the United Nations in 1987, Japan and Morocco proposed that the final decade of the twentieth century (1990–1999) be designated an "International Decade for Natural Disaster Reduction," and this resolution was unanimously adopted. This can be seen as the first time that the international community formally recognized disasters as shared challenges that require globally coordinated actions. The years following the adoption of this resolution witnessed the frequent occurrence of large-scale disasters around the world and, along with the progression of debates on climate change and sustainable development, the importance of disasters and disaster reduction as a global issue increased. In 1994, the mid-year of the International Decade for Natural Disaster Reduction, the first United Nations World Conference on Natural Disaster Reduction was held in Yokohama, Japan. At this meeting, the Yokohama Strategy, a series of principles and plans for action to promote disaster reduction, was adopted. In order to extend the worldwide efforts initiated by the International Decade for Natural Disaster Reduction further into the future, in 2000 the International Strategy for Disaster Reduction (UNISDR) was established within the United Nations. The UNISDR aims to establish and coordinate international cooperation for disaster response and to support disaster-reduction measures around the world. The World Summit on Sustainable Development held in Johannesburg, South Africa in 2002 marks the first time that leaders from around the world recognized natural disasters as a threat to humanity and issued a summit declaration emphasizing the need to promote disaster reduction. These events, and the establishment of international organizations in the latter part of the twentieth century, clearly illustrate the increasing global-scale institutionalization of post-disaster support and disaster reduction.

Worldwide recognition of disasters advanced even further following the turn of the century. Nevertheless, it is critical to recognize that this movement does not simply mean the "up-scaling" of disaster response and support. Following the unprecedented devastation caused by the Indian Ocean tsunami disaster of December 2004, government delegations from 168 countries participated in the 2nd United Nations World Conference on Disaster Reduction in Kobe, Japan in January 2005. The importance of international cooperation to promote disaster reduction was strongly advocated by delegates to this conference and the Hyogo Framework for Action (HFA) 2005–2015 was formulated in order to establish guidelines for action to create a more disaster-resilient world. The HFA outlines basic and comprehensive policies for promoting disaster reduction until the year 2015 and establishes the goals and priorities shown in Figure 2.2.

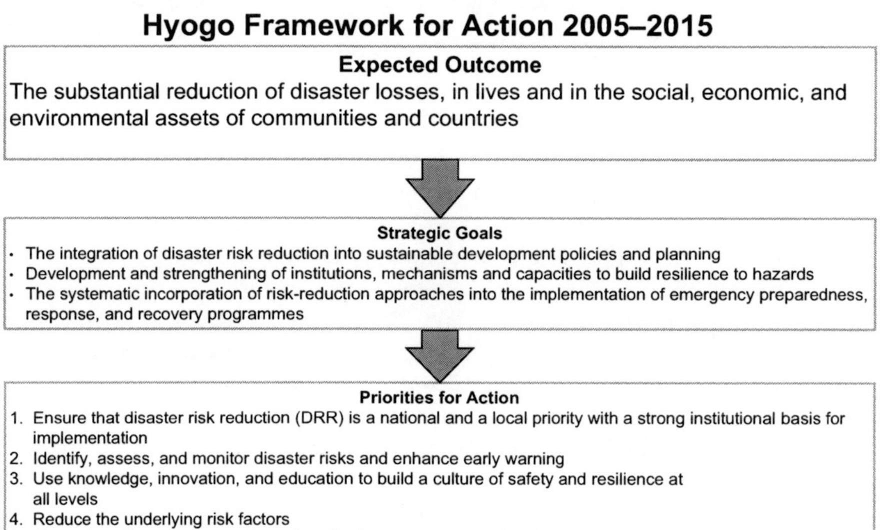

Figure 2.2 Outcome, goals, and priorities of the Hyogo Framework for Action 2005–2015.

One of the novel features of the HFA is the recognition of the long-term effects of disasters on social and economic development, rather than merely their short-term impacts on human lives and physical infrastructure. In other words, disaster risk reduction had been seen primarily as a humanitarian issue, rather than an essential element of sustainable development. Despite this important change, in reality most countries still continue to prioritize post-disaster response and reconstruction, rather than pre-disaster resilience, in their disaster-related policies.

Although national and community planning for disaster preparedness based on the HFA bas been implemented around the world, as evidenced in the cases of Hurricane Katrina in the United States, the Sichuan earthquake in China, the Haitian earthquake, and the Great East Japan Earthquake Disaster, large-scale disasters continue to occur in both developed and developing countries. In 2015, the final year of the HFA and four years after the Great East Japan Earthquake Disaster, the 3rd World Conference on Disaster Risk Reduction (WCDRR) was held in Sendai, Miyagi Prefecture, on March 14–18. More than 80,000 people from 187 countries attended the conference and its side events. In the conference, the Sendai Framework for Disaster Risk Reduction (SFDRR) 2015–2030 was adapted as the new long-term guideline for disaster risk reduction, succeeding the HFA.

The SFDRR marked a significant shift from the HFA and other earlier frameworks in that it explicitly embraces natural disasters as well as man-made and technical disasters, as defined in its scope:

> The present Framework will apply to the risk of small-scale and large-scale, frequent and infrequent, sudden and slow-onset disasters caused by natural or man-made hazards, as well as related environmental, technological and biological hazards and risks. It aims to guide the multihazard management of disaster risk in development at all levels as well as within and across all sectors.
>
> (UNISDR 2015, 11)

In other words, the international community now recognizes the necessity of risk reduction for cascading disasters such as a nuclear accident consequent to a natural disaster. While the HFA had aimed at "reduction of disaster losses," "reduction of disaster risk" was explicitly included in the SFDRR, emphasizing preparation for pre-disaster investment for reducing disaster risk (Figure 2.3).

The SFDRR also emphasizes that both developed and developing countries should aim to reduce disaster risks through stakeholders' capacity-building, in addition to pre-disaster investment in risk reduction. This means that policymakers must make disaster risk reduction an essential part of their development policies and programs. The SFDRR recognizes that some disaster damages are virtually irrecoverable, that a single disaster could wipe out all development efforts in an instant, and that its adverse effect on development and poverty reduction may last for a long time. Lack of appropriate preparation for DRR threatens basic human security, especially in the case of complex natural- and human-induced disasters, such as that experienced in Fukushima.

Sendai Framework for Disaster Risk Reduction 2015–2030

Expected Outcome

The substantial reduction of disaster risk and losses in lives, livelihoods, and health and in the economic, physical, social, cultural and environmental assets of persons, businesses, communities, and countries

Goal

Prevent new and reduce existing disaster risk through the implementation of integrated and inclusive economic, structural, legal, social, health, cultural, educational, environmental, technological, political, and institutional measures that prevent and reduce hazard exposure and vulnerability to disaster, increase preparedness for response and recovery, and thus strengthen resilience

Priorities for Action

1. Understanding disaster risk
2. Strengthening disaster risk governance
3. Investing in disaster risk reduction for resilience
4. Enhancing disaster preparedness for effective response, and **Build Back Better** in recovery, rehabilitation, and reconstruction

Figure 2.3 Outcome, goals, and priorities of the Sendai Framework for Disaster Risk Reduction 2015–2030.

The "Build Back Better" approach and international cooperation

The SFDRR guidelines highlight the notion of "Build Back Better" (BBB). BBB is generally understood as a process of creating more resilient communities and nations than was the case before a disaster through the implementation of well-balanced disaster risk-reduction measures, including the physical restoration of infrastructure; revitalization of livelihood, economy, and industry; and restoration of the local culture and environment (Matsumaru *et al.* 2015). The term was originally coined after the Indian Ocean tsunami of 2004. Amid reconstruction operations in the heavily devastated Aceh region, the slogan "implementing building back better at every opportunity" was used to emphasize the construction of better social infrastructure, community-led reconstruction, the empowerment of vulnerable people, and strengthening the resilience of communities, and intensive operations were carried out. Additionally, it was urged not only to build physically more robust structures to protect communities from disasters but also to establish urban planning reflective of disaster considerations and to build social capital through frameworks that allow every stakeholder to get involved. As a result, in a region that had experienced conflicts for the 30 years prior to the disaster and that had long been the least developed region in Indonesia, high economic growth (annual average growth of 4.4 per cent from 2005 to 2011) and population increases (from 178,000 in 2005 to 245,000 in 2011) were experienced (BRR 2009; BAPPEDA Aceh 2011).

In the past decade the concept of BBB has become widely used around the world, as seen in the adoption of the term in the SDFRR. There are several key reasons for the concept's widespread popularity. The first is a change in the perception of the cost-effectiveness of pre-disaster preventive measures. In the past, most donor agencies dealing with disasters placed emphasis on post-disaster response, relief, and recovery; this was not only based on humanitarian considerations, but was also because expenditures for disaster-prevention equipment and facilities were thought to be inefficient allocations, since there can be no knowledge of when a natural disaster will occur (UNDP 2008). However, in recent years, as a result of the growing frequency of large-scale disasters, the notion that it is more logical to take preventive measures against potential disasters than to conduct recovery efforts after a disaster has become widely accepted. According to the United Nations Development Programme, preparedness saves more lives than response, and "every dollar spent reducing people's vulnerability to disasters saves around seven dollars in economic losses" (UNDP 2012, 1).

Second, there is a growing recognition of the critical linkages between disasters and poverty. Particularly in developing countries, immediate post-disaster relief efforts, although still essential, may have a limited long-term contribution to the resilience of the disaster's victims. As some natural disasters tend to recur in the same locations, victims of disasters are often trapped in a spiral of poverty. Third, two large-scale changes in structural parameters—climate change and global urbanization—demand that we consider whether we may be entering into a qualitatively different "world of disasters." As the frequency and magnitude of disasters appear only to increase with these changes, it seems increasingly difficult even to restore pre-disaster conditions if we stick with conventional post-disaster responses.

For all these reasons, the idea of BBB is likely to stay with us for the foreseeable future. That is, recovery and reconstruction after a natural disaster should be carried out with an image of the future livelihoods and industries of the area in mind, and it is also imperative to actively conceptualize the experience of a disaster as an opportunity to rethink urban planning and development policy, as well as to prepare for the next disaster.

The SFDRR proposes that donors change their international cooperation priorities from emergency response and recovery to support for pre-disaster preparation measures, including prior investment, and the promotion of "pre-disaster reconstruction," a form of proactive community-based urban planning that recognizes the potential for disasters and seeks to minimize damages before a disaster occurs. However, most donors and recipient countries continue to adhere to the "old" ideology, regarding response and recovery as disaster-management priorities. This is due largely to the fact most countries regard DRR as a secondary priority and economic development as the first priority. This is also related to the fact that the effects of DRR investment are not always visible unless a disaster actually occurs (Nagatomo 2015). Thus, achieving the objectives of the SFDRR requires a paradigm shift. Toward achieving such a shift, Japan, the largest bilateral donor of disaster aid, expressed its intention to support implementation of the SFDRR by focusing on prior investment and BBB as a priority, and declared the Sendai Initiative as its commitment to DRR assistance (Ministry of Foreign Affairs 2015). The concept of BBB has been particularly strongly supported by countries affected by large-scale disasters. Indonesia applied it to reconstruction following the Indian Ocean tsunami of 2004, and the Philippines entitled its reconstruction plan after Typhoon Yolanda "Build Back Better" (Government of Philippines 2013). The concept has thus diffused widely from these countries that experienced severe disasters.

Conclusion

This chapter has outlined the evolution of institutions and practices for international cooperation in disaster risk reduction. International relief efforts and support activities after the Great East Japan Earthquake Disaster went much more smoothly and were better coordinated than those after the Hanshin earthquake disaster, closely reflecting this institutional evolution. At the same time, the reconstruction process in place after the most recent Japanese disaster also mirrors the problems and remaining challenges in DRR addressed in this chapter, namely the recognition and implementation of the Build Back Better concepts. For example, operations to move the afflicted communities to higher ground and to build new seawall defenses are being advanced throughout the coastal areas of Tohoku; yet, we are seeing cases in which citizen participation in the decision-making process has been often inadequate, or in which a failure to collect various opinions has resulted in delays in reconstruction operations. While the SFDRR asserts that disaster risk reduction should be placed within the mainstream of development planning, there remains a need to expand efforts to reduce disaster risks as part of the reconstruction projects that have followed the Great East Japan Earthquake.

The 2015 Goroka earthquake in Nepal indicated progress in the international coordination of emergency response and humanitarian aid, in which the importance of the Build Back Better approach was also clearly recognized (Government of Nepal 2015). Additional mechanisms to mitigate disaster risks, such as disaster insurance and stand-by credit schemes, were also developed (World Bank 2015), and those mechanisms improve response and recovery from disasters. On the other hand, there is criticism that such financial-compensation mechanisms affect governments' and communities' motivations for DRR investment. In order to prevent such moral hazards and improve the recipients' disaster-management systems, technical and financial support should be provided simultaneously (CTI Engineering International 2010).

We should question whether reconstruction in Fukushima is aiming to build back better than before the disaster. Unfortunately, it is difficult to answer that question in the affirmative at this point. This is in part because Build Back Better emerged, and has been applied thus far, in contexts where communities stricken by large-scale *natural* disasters have sought to build resilience to future disasters.

Challenges presented by the Fukushima nuclear disaster, including the loss of land due to radioactive contamination, the division of families and communities, and the psychological effects of radiation—unlike the challenges presented by earthquakes or tsunami—persist long after the onset of the disaster. There is also an issue of accountability. Since the perpetrator of this accident is clearly known, emphasis has been placed primarily on how to compensate for damages, while reconstruction focused on improving quality of life in the area has been arguably secondary. Finally, current reconstruction plans in Fukushima place a strong focus on environmental restoration and industrial revitalization; comparatively less attention has been given to planning for future disasters (Fukushima Prefecture 2012). The decommissioning of all nuclear reactors, a stated goal of the prefecture, might be the best way to prevent another nuclear disaster, but an exclusive focus on nuclear disasters may leave the prefecture unprepared for other disasters. All of this implies that building back better in Fukushima will likely require novel thinking and creative actions, and that the lessons and hardships, as well as the successes and failures, of the recovery process in Fukushima must be widely shared around the world.

Acknowledgment

This work was supported by JSPS KAKENHI Grant Numbers 25220403 and 26560181.

References

BAPPEDA Aceh. 2011. *Statistik Daerah Provinsi Aceh 2011* [Statistics of Aceh Province 2011]. Banda Aceh: Aceh Province. [In Indonesian]

Cabinet Office of Japan. 2011. *White Paper on Disaster Management 2011*. Tokyo: Cabinet Office of Japan.

CTI Engineering International and Nippon Koei. 2010. *The Preparatory Study for Sector Loan on Disaster Risk Management in the Republic of the Philippines: Final Report*. Tokyo: Japan International Cooperation Agency.

Executing Agency for Rehabilitation and Reconstruction (BRR). 2009. *10 Management Lessons for Host Governments Coordinating Post-disaster Reconstruction*. Jakarta: Executing Agency for Rehabilitation and Reconstruction.

Fujimoto, Noritsugu. 2014. Haichi Ojishinto Makurobaransu [The Haiti Earthquake and Macro Balance]. In *Higashinihon Daishinsaikarano Hukkyu/Hukko to Kokusaihikaku*. [Recovery and Reconstruction from Great East Japan Earthquake and International Comparison], edited by Fukushima University International Disaster Recovery Study Team, pp. 241–260. Tokyo: Hassaku-sha. [In Japanese]

Fukushima Prefecture. 2012. *Plan for Revitalization in Fukushima Prefecture (Second version)*. Accessed January 5, 2016. www.pref.fukushima.lg.jp/download/1/plan_for_revitalization2_outline.pdf.

Government of Nepal, National Planning Commission. 2015. *Nepal Earthquake 2015 Post Disaster Needs Assessment vol. 1 Key Findings*. Kathmandu: National Planning Commission.

Government of Philippines, National Economic Development Authority. 2013. *Reconstruction Assistance on Yolanda Build Back Better*. Accessed June 30, 2016. www.gov.ph/downloads/2013/12dec/20131216-RAY.pdf.

Herod, Andrew. 2011. What Does the 2011 Japanese Tsunami Tell us About the Nature of the Global Economy? *Social & Cultural Geography* 12(8): 829–837.

Katayama, Yu. 2013. Higashinihon Daishinsaijino Kokusai Kinkyuu Shien Ukeireto Gaimusho [International emergency assistance during the Great East Japan Earthquake and the Ministry of Foreign Affairs]. *Journal of International Cooperation* 20(2): 45–73. [In Japanese]

Kawada, Yoshiaki. 2003. *Bousaito Kaihatsu – Shakaino Bousairyokuno Koujouwo Mezashite* [Disaster Management and Development—To Improve Social Capacity]. Tokyo: Japan International Cooperation Agency. [In Japanese]

Matsumaru, Ryo and Kimio Takeya. 2015. Recovery and reconstruction: an opportunity for sustainable growth through "Build Back Better." In *Disaster Risk Reduction for Economic Growth and Livelihood—Investing in Resilience and Development*, edited by Ian Davis, pp. 139–160. New York: Routledge.

Ministry of Foreign Affairs of Japan. 2014. *ODA chotto iihanashi No.1 Kanto Daishinsai wa New York Times shi no top news* [ODA Episode No.1 The Great Kanto Earthquake as a headline of the *New York Times*]. Accessed November 11, 2015. www.mofa.go.jp/mofaj/gaiko/oda/hanashi/story/1_1.html. [In Japanese]

Ministry of Foreign Affairs of Japan. 2015. *Sendai Cooperation Initiative for Disaster Risk Reduction*. Accessed June 30, 2016. www.mofa.go.jp/mofaj/files/000070664.pdf.

Nagatomo, Noriaki, Otsuki, Eiji and Hirano, Junichi. 2015. Economic analysis of investment in DRR measures. In *Disaster Risk Reduction for Economic Growth and Livelihood—Investing in Resilience and Development*, edited by Ian Davis, pp. 139–160. New York: Routledge.

New Zealand Police. 2012. *List of deceased, data as at 9 February 2012*. Accessed June 30, 2016. www.police.govt.nz/major-events/previous-major-events/christchurch-earthquake/list-deceased.

Nishikawa, Satoru. 1996. Hanshin Awaji Daishinsaini Mirareta Kokusai Kyuuen Katsudouno Misumacchi [The mismatch of international relief aid in the Hanshin Awaji Earthquake Disaster]. *Journal of the Institute of Social Safety Science* 6: 261–268. [In Japanese]

Ranghieri, Federica and Mikio Ishiwatari. 2014. *Learning from Megadisasters—Lessons from the Great East Japan Earthquake*. Washington, DC: World Bank.

United Nations Development Programme (UNDP). 2008. *Human Development Report 2007/2008*. New York: UNDP.

United Nations Development Plan (UNDP). 2012. *Putting Resilience at the Heart of Development: Investing in Prevention and Resilient Recovery*. New York: UNDP.

United Nations International Strategy for Disaster Reduction (UNISDR). 2012. *The Economic and Human Impact of Disasters in the Last 12 Years*. Accessed June 30, 2016. www.unisdr.org/files/25831_20120318disaster20002011v3.pdf.

United Nations International Strategy for Disaster Reduction (UNISDR). 2013. *Global Assessment Report on Disaster Risk Reduction 2013*. Accessed June 30, 2016. www.preventionweb.net/english/hyogo/gar/2013/en/home/index.html.

United Nations Office for Disaster Risk Reduction (UNISDR). 2015. *Sendai Framework for Disaster Risk Reduction 2015–2030*. Geneva: UNISDR. Accessed January 5, 2016. www.unisdr.org/files/43291_sendaiframeworkfordrren.pdf.

United Nations University, Institute for Environment Human Security (UNU-EHS). 2014. *World Risk Report 2014*. Bonn: UNU-EHS and Alliance Development Works.

World Bank, Global Facility for Disaster Risk Reduction. 2015. *Disaster Risk Financing and Insurance*. Accessed November 11, 2015. www.gfdrr.org/disaster-risk-financing-and-insurance.

Yanagisawa, Kae. 2013. *Daisaigaini Tachimukau Sekaito Nihon Saigaito Kokusaikyouryoku* [Japan and the World Confronting Disaster: Disasters and International Relations]. Tokyo: Sakai Printing. [In Japanese]

3 Challenges of just rebuilding

Case studies of Iitate Village and Tomioka Town, Fukushima Prefecture

Akihiko Sato

Introduction

The unprecedented scale of the nuclear accident at the TEPCO Fukushima Daiichi Nuclear Power Plant has forced victims into a life of evacuation with no end in sight. The prolonged evacuation has led to a shift in general public sentiment, and the particular sentiment of evacuee-receiving communities, from warm sympathy to victim-blaming, resulting in mounting stress for the evacuees. In addition, with anxiety over the crippled reactors still unresolved, an increasing number of people have given up on returning to their former homes—a fact clearly indicative of a major change in the attitudes of victims. However, even five years after the disaster, studies exploring the conditions, problems, and changes facing disaster victims remain far from adequate.

Since the 1950s, after major environmental pollution causing Minamata disease and Yokkaichi asthma, Japanese social scientists have accumulated a large volume of research on environmental pollution and societies (Dunlap 1995; Hasegawa 2004). A series of studies by Nobuko Iijima on the social factors and structures (e.g. the role of perpetrators who cause pollution and the government's role as their defender) that exacerbate the effects of environmental pollution to life and health (Iijima 1984), and Harutoshi Funabashi's analysis of "derivative perpetrators," are of particular theoretical importance (Funabashi 1999). However, these existing theoretical perspectives, which focused mostly on "conventional" pollutants, cannot simply be applied to the case of the nuclear accident. The main reason for this is that nuclear-disaster evacuees are confronting a situation in which the direct physical damage to individual life and health (in particular the effects of radiation) is difficult to discern in the short term, while a diverse range of indirect effects—break-up of families, intergenerational conflicts, loss of community—are simultaneously taking place, threatening the very foundations of victims' inhabitance, life, and health. Thus, the relationship between perpetrator and victims used as a basis by previous research on environmental pollution is entirely different in the case of the nuclear disaster. Accordingly, the following discussion reviews studies on evacuation from the nuclear disaster-afflicted areas and on the damages that resulted from the incident in order to identify the state and limits of current research.

One body of research on nuclear-disaster evacuation and damages has focused on the local governments afflicted by the disaster. Studies within this area of research have focused mainly on such themes as the situations confronting local governments as a result of the disaster, the sequences of evacuation, the process of reorganizing administrative functions, and the policy process before and after the earthquake disaster (e.g. Kawazoe and Urano 2011; Ohashi and Takaki 2012; Kanno and Takaki 2012; Yanagisawa 2012; Sato 2012, 2013). A second body of research has dealt with the compulsory and voluntary evacuations that followed the nuclear-power accident (e.g. Yamashita and Kainuma 2012). A third group of studies focuses on victims who have evacuated beyond Fukushima Prefecture and the support policies adopted by local governments receiving these evacuees (e.g. Matsui 2012; Takaki *et al.* 2012). All these studies provide important insights into the conditions and challenges faced by the evacuees. Five years have now passed since the nuclear accident; the afflicted communities' situations have radically transformed and are still evolving. Accordingly, it is vital that the scientific community continues to observe the situation and compile detailed records on the affected areas and victims.

While the above groups of research have adopted a local case-study approach on the disaster-afflicted local governments and individuals, a fourth group of studies has quantitatively analyzed the conditions of evacuees across multiple localities (e.g. Tamba 2012; Imai 2011a, 2011b, 2012). While these studies identify evacuees' conditions after the accident, and therefore provide an essential baseline for conceptualizing reconstruction policies, the representativeness of the sample populations from which these studies draw has been questioned. A fifth and final body of research deals with issues of compensation for damages resulting from the nuclear disaster and evacuation (Yokemoto 2013). This body of research theorizes the loss of hometown—that is, the loss of tangible possessions as well as intangible and irreplaceable relations—as the most basic form of damage shared by evacuees, and explores whether and how it may be possible to compensate victims for such damages. These studies analyze the prospects for compensation and criticize the trivialization of the damage caused to victims' lives on the part of the powerful perpetrators of the accident (i.e. TEPCO and the national government), as well as the way in which damages have been addressed without actually assigning the perpetrators any responsibility for the accident (Yokemoto 2013).

The above discussion has focused mainly on trends in research on nuclear evacuation and other related issues. As can be seen, the majority of these studies focus on very specific problems confronting individual victims and communities, while the actual challenges facing evacuees are of a compounding nature, one conditioned by social and political dynamics operating at multiple different spatial scales. This chapter attempts to unravel the cross-scalar dynamics underlying the experiences of evacuees. Accordingly, this chapter will look at the types of problems that evacuees experienced after the disaster and how they have changed, and will furthermore examine how these problems are intertwined with social and political structures operating at multiple scales. This chapter will examine case studies of Iitate Village and Tomioka Town, two municipalities from which numerous evacuees originated, to develop some concrete suggestions to achieve the reconstruction of victims' lives and livelihoods.

Civil society and multi-structures as key to conceptualizing reconstruction

In contemporary Japan, the term "society" (*shakai*) has double connotations. It can mean both "civil society," as in contemporary western social-scientific literature, and "community" (hence, "regional society" is more or less used synonymously with "regional community"). The former emphasizes the nature of society in relation to other sectors, such as the state (i.e. a bimodal model of society–state) or the state and economy (i.e. a trinomial model of society–state–economy). The latter connotation, by contrast, has a more "internal" focus and generally frames a society as an arena in which the concrete lives of individuals, households, and organizations unfold. While the latter perspective is essential in understanding the events that followed the earthquake disaster in Fukushima, I wish to call attention to the importance of complementing this "community" focus with a broader perspective on society as intertwined with state and economy.

Although there are many other, often more complex or highly debated, models of society that might be drawn upon, for the purpose of my argument it suffices to point out, drawing on the trinomial model, that the above-noted three spheres—society, state, economy—mutually influence one another. For example, the state regulates our actions through laws and policies. On the economic side, we receive compensation for selling our labor power, which then supports the economy through our consumption. In this way, our lives (society) are always conditioned by the state and the economy. The problem here is that economy can begin to trump society, as the state and economy collude to advance the "colonization of the lifeworld" (Habermas 1985) and our lives become increasingly ruled by market principles (Shinohara 2004). For example, the development of electrical engineering, nuclear technology, and genetic science, and the various laws and policies through which these technologies unfold, have all come to have a large influence on various aspects of our lives, including convenience and comfort.

Put another way, the economy has come to play the role of the *subpolitical*. The term "subpolitics" has been used by scholars such as Beck (1997) to describe the (growing) tendency for actual politics (decision-making, negotiation, legitimization, etc.) that shape society to take place outside rule-based political systems (hence subpolitics is *sub(system)politics*) (Beck 1997). In the context of this chapter, although defects in the nuclear reactors had been repeatedly pointed out prior to the accident, the state accepted TEPCO's claims of safety—that is, the state followed along with the subpoliticized economic sector, which was assumed to have the necessary expertise and knowledge to make such claims. It is for this reason that the defects leading to the nuclear accident were not sufficiently remedied. Under such circumstances, it is necessary for civil society to monitor the political and ethical aspects of such subpolitics (i.e. the actions of companies and the decisions of science) (Shinohara 2004). In particular, such monitoring by civil society typically concerns the realms of welfare, environment, medicine, and other matters that involve *local* lives and events. In Fukushima, then, it is imperative for civil society to monitor the validity and reliability of data pertaining to the risks and safe management of radiation. Unfortunately, such monitoring is not currently being conducted in Fukushima to any satisfactory degree.

Another problem is that while the siting of the Fukushima Daiichi Nuclear Power Plant was in part a component of a broad national-level energy policy, for the local community the formation of an industrial structure revolving entirely around TEPCO was a local phenomenon. Accordingly, there are multiple subpolitical dynamics at work at different spatial scales.

Finally, the capacity for civil society to monitor the subpoliticized economic sector is integrally related to the issues of local self-governance, local political autonomy, and administrative decentralization/devolution. Indeed, it is precisely the inadequate functioning of local self-governance that has produced various fissures between victims' views, local governments' intentions regarding reconstruction, and the state's views on the issue of nuclear restart and energy policy after the disaster. Here it can be said that several seemingly discrete problems—the issues of subpolitics, state and region, and the problems of local autonomy—all coalesce. In what follows, we will look at the cases of Iitate, a village designated a "deliberate evacuation area," and Tomioka, a town forcibly evacuated after the disaster and later designated a "restricted area," in order to explore the above-noted issues.

The state of affairs after the earthquake disaster: the case of Iitate Village[1]

Around 1980, the village of Iitate began to pursue unique forms of community-based planning and revitalization, or *machi-zukuri* (Sorensen and Funck 2007). While these efforts at revitalization were pursued relatively independently and autonomously in the early years, cracks began to show from the latter half of the 1990s (Sato 2013). Notions of "new public management" rapidly spread programs of administrative reform and financial rationalization throughout Japan during this period. Amid reductions in local-government staff and budgets, local municipalities' reliance on state subsidies increased, and the capacity for local governments to pursue independent policies weakened. The result is that even faced with the massive disaster of a nuclear accident, local governments are forced to follow along with state policies and guidelines, rather than listening to the wishes and demands of their communities.

Said somewhat differently, a number of issues that were slowly eroding the foundations of local governance prior to the accident suddenly ruptured these foundations wide open in the processes of forced evacuation, evacuee return, and reconstruction that followed the nuclear accident. Critically, the conditions for such rupture are latent in almost any locality in contemporary Japan. Indeed, we continue to accept, willingly or unwillingly, the rationale and workings of the regime based on the state–local relationship formed after the 1990s in Japan, which led to the historic disaster in March 2011. Let us pursue this point further below.

The conditions of planned evacuation

On April 22, 2011, Iitate Village was designated a "deliberate evacuation area"; four months later, evacuation of the village was completed. As of January 1, 2016, the post-evacuation population of the village consisted of 6,729 evacuees and 3,042 households, according to the Iitate municipal office (Iitate Village 2016). At that time, approximately 90 per cent of the town's population were living in evacuation within a one-hour commute of the town office. The separation of households—in both an emotional and a geographic sense—has greatly expanded as a result of the disaster and prolonged evacuation, with the total number of households now 180 per cent of that which it was before the disaster. The fragmentation of households is likely caused by such occurrences as prolonged loneliness leading the elderly to move to temporary housing to find companionship; families separating as children move to new schools; and one parent moving to a different place due to work considerations.

As the evacuation directive of Iitate Village is expected to be lifted in March 2017, approximately 83 per cent of the population must weigh the pros and cons of return and eventually make a decision as to whether or not they will return to their homes.

Distrust of government spread through dialogue with citizens

In October of 2011, the village announced its "Iitate Reconstruction Plan," and throughout December meetings were held at prominent evacuation destinations. The main purposes of these meetings were 1) to inform villagers of basic policies for reconstruction, and 2) to take account of villagers' opinions and wishes so that these results could later be incorporated into reconstruction planning.

However, contrary to the aims of the village government, the villagers assembled at these meetings roundly criticized the mayor and assemblymen of the village. These criticisms centered mainly on three issues: 1) concerns over the feasibility of decontamination; 2) demands to use part of the more than 300 billion yen budgeted for decontamination to enhance the reconstruction of daily life and business; and 3) demands for collective relocation in the near term. In response to each of these concerns and criticisms, the mayor and executive officers of the village explained that they were presently unable to take these views into consideration because it was the national government that designated the "deliberate evacuation areas." Accordingly, all responsive measures to be carried out within these areas, such as decontamination, would be the sole responsibility of the national government. However, the same village officers also admitted, somewhat contradictorily, that the national government would not act without pressure from the village and that, even in the case of decontamination, the national government would only begin to act once the village put forth the appropriate expenses. Essentially, while the village officers admitted the importance and effectiveness of village agency, they were also unwilling to challenge the national government. They also argued that the national government could only provide funding because the target of operations was decontamination. It would be impossible, they said, for the national government to use tax money to compensate individual victims. Moreover, since negotiations with the national government are currently focused on implementing decontamination as an essential step toward returning residents to the evacuated villages, it is not possible to demand that the national government support plans to relocate (Sato 2012, 119–120; Sato 2013, 40–41). In short, the village government was unwilling to question the validity of the larger framework for reconstruction planning laid out by the national government.

Villagers' responses to this explanation are well encapsulated by one villager who stated: "we understand the need to value and care for the 'village,' but we want you to also think about the 'lives' of villagers." This unilateral cutting-off of dialogue by the village office eventually led to a decrease in the number of participants at the meetings. In the beginning, relatively young heads of household (30–60 years old, who were previously typically too busy to attend these meetings) and their spouses (which is also quite unusual), as well as older individuals who had rarely participated in these meetings before the accident, were prominent among the attendees. Yet, with every meeting, the number of participants dropped.

One resident who refused to attend these meetings stated: "they only inform us of things that are already decided and would never listen to our opinions" and "if that's the case, then participating is a waste of time." The result of the continuation and accumulation of this conflict has been a decrease in the quality and quantity of opportunities for dialogue between residents and local officials. The local government continues to make numerous important decisions without any compromise from either side, which further divides citizens and local officials, fosters distrust, and has even led some residents to change their residence registry and leave the village behind. This problem is not limited to Iitate Village, but rather seems to be shared by all of the nuclear disaster-afflicted areas. One of the reasons, as outlined above, is the problematic relationship between the state and localities that has emerged since the 1990s in Japan. Municipal governments have been forcibly restricted to following national government policies and, accordingly, municipalities in the disaster-afflicted areas have been unable to serve their community's interest.

Efforts to rebuild the disaster-afflicted areas and the lives of victims

The dissatisfaction with administrators observed in Iitate Village and the reduction of trust in the politico-governmental apparatus seem to be issues shared by many of the nuclear disaster-afflicted municipalities. However, in Tomioka Town, a residents' self-help association has been at the center of efforts to transform this situation by putting the perspectives of victims back onto the agenda and by infusing this perspective into the development of reconstruction policies.

Outline of a town meeting

The Tomioka Children's Future Network (TCFN), a self-help association composed of and led by the citizens of Tomioka, has been the key agent behind the implementation of a series of town meetings. Although they are termed "town meetings," these meetings are actually mobile meetings held in venues across the country, thus enabling dispersed evacuees from Tomioka to assemble and discuss the shared but separately experienced issues and difficulties with which they are grappling. These repeated meetings have brought into focus present issues and future tasks and have thus begun to provide a coherent and synergistic foundation for appeals to administrators and government officials (Tomioka Children's Future Network 2013). As of April 2013, eight mobile town meetings had been held around the country in large cities such as Koriyama, Iwaki, Nagaoka, Tochigi, Yokohama, and Tokyo. Reflecting back on the town meetings and the TCFN's other activities, leader and representative Ichimura suggests that "this is an effort to join together with other townspeople to identify problems and tasks such that these views can then be convincingly presented to the national government and administrators." According to Ichimura, the fact that the current plight of evacuees "has not been correctly understood by the government or by the public (that can influence politics and administrative services)" has been the catalyst for these town meetings (Tomioka Children's Future Network 2013, 13).

The problem structure brought into focus by the voices of evacuees

Through the series of TCFN meetings it has become apparent that the problems with which evacuees are dealing span a wide range of scales, from personal to family, from community to local-government issues. First, on a very personal level (Figure 3.1), evacuees feel deprived of their lives, their ways of living, and their pride. The sense of deprivation is fueled by the challenges they face at evacuation destinations, such as finding suitable work and receiving a good education. In addition, they feel very frustrated by the fact that such a sense of deprivation is unlikely to be understood or supported by public opinion. At the family level, many of the evacuated residents of Tomioka identify the separation of family members and fragmentation of households as a serious problem. At the community level, evacuees identify the loss of social ties with neighbors, friends, and close confidants that they had built up over many years. Some also frame such loss as the deprivation of temporal rhythms (e.g. those shared by mothers of the community) and local environment (e.g. for nurturing child development) that constituted life in Tomioka. At the level of the local town, members are torn between the ideas of returning or not returning to Tomioka, even though a desire

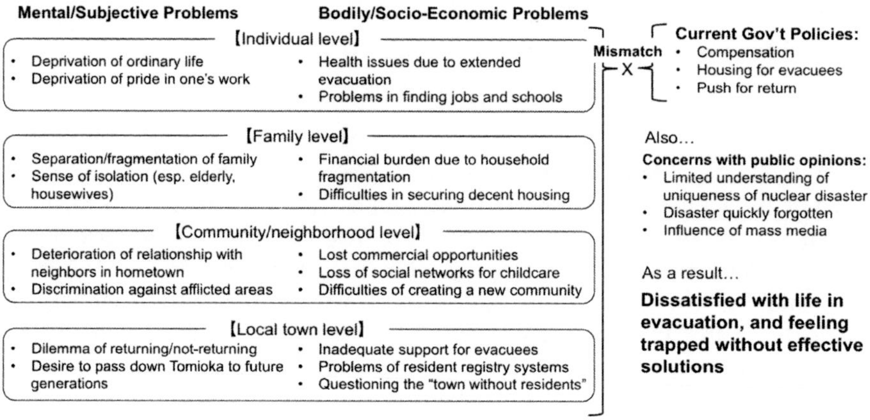

Figure 3.1 Structure of the problems faced by evacuees, revealed through town meetings in Tomioka.

Source: Author.

to pursue place-rebuilding efforts that would enable a palpable sense of Tomioka to be passed down to future generations is a shared feeling that crosses any generation gap.

These problems, though felt and experienced at different scales, are thoroughly intertwined. For example, problems in children's mental and physical health (personal issues) may be caused and amplified by the problems experienced by their family members, such as the isolation and loneliness of mothers and elderly caretakers in the evacuation destinations (community issues). This also means that the mitigation of personal problems may be aided by conditions at other scales. For example, children who have trouble fitting in at schools at evacuation sites, or parents who experience exclusion due to their evacuee status, may stay resilient because of their ties to their home communities.

Nevertheless, evacuees are strongly dissatisfied and concerned with the conditions brought about by having lost the everyday lives they once took for granted, and many have raised concrete demands that would allow them to recover normal life. Some of these voices were responded to by government policies on subsidized housing, free use of expressways for victims, and the contemplation of a double-residence registry system. However, policies such as the "temporary town" do not adequately address the problems and tasks outlined above. Indeed, the current policies are mostly ad-hoc, special measures that are reviewed and renewed annually—far from fundamental institutional reforms. Additionally, public opinion has a powerful influence on politics and administrators, and such opinions are strongly shaped by mass media. There is concern among evacuees over the quality of the media, which tends to focus on sensational events and to overly simplify otherwise hugely complex issues faced by the evacuees. Moreover, as Fukushima gets less and less media attention itself, more than a few evacuees are highly worried that any public support for the plight of evacuees will wither in the future.

Issues and future tasks for rebuilding localities and lives

When taking a broader view of the structure of the problems that have enveloped evacuation life, as we did in the previous section, it becomes clear that nearly all of these problems are related to the current systems' and policies' inability to adequately respond to the plight of victims. Over five years have slipped by, and victims continue to suffer from various hardships and daily conflicts. For some of these victims, the opportunity to briefly visit their homes in the restricted areas changed from being part of the process of returning to a final opportunity to say farewell to their homes and hometowns. For all evacuees, the question of whether they can continue to be citizens of Tomioka weighs heavily. It will not be surprising if many evacuees, during the course of finding housing and employment and rebuilding their lives at their evacuation destinations, opt to become residents of the latter towns for the long term. The issue, however, is that such a decision to change one's residence registry could lead to loss of the compensation and care for future health risks that was supposed to be granted to evacuees, as well as termination of the invaluable social relations they would leave behind in Tomioka. Moreover, the decision to permanently relocate would extinguish citizens' ability to rely on support from TEPCO or the national government, leaving evacuees to fend completely for themselves.

The Science Council of Japan (2013, 7) points out that failure to improve the harsh conditions of evacuee life and chances for restoration of previous conditions would, when combined with already existing trends toward depopulation, threaten the future survival and very existence of the local governments and communities of the nuclear disaster-afflicted areas. If only 20 per cent of evacuees return to the evacuated areas near the nuclear plant, then there is a strong possibility that local governments would need to merge in the future in order to survive. However, if existing local governments were to disappear, the result would be the eradication of the very political channels through which residents have attempted to make known and resolve the many issues they have confronted since the nuclear accident.

Conclusions: toward rebuilding localities, lives, and civil society

These case studies have revealed issues and problems in rebuilding the lives of the disaster victims and evacuees which are intricately related to the challenges of empowering civil society. Based on the findings, I consider several key tasks for the future below.

Securing procedural justice

In our case study of Iitate Village, we saw how the residents' meetings that were supposed to serve as venues for constructive dialogue were transformed into sites for contentious debate that furthered distrust between government and residents. This was caused by inadequate facilitation of *two-way* dialogue, which would have been essential for securing mutual understanding between residents and government. This eventually led to the roll-back of opportunities for *any* dialogue. While speech is an important element of procedural justice, what is most important is to guarantee two-way dialogue. Even if dissatisfaction and criticism develop, the provision of a valid explanation for why a party's demands cannot be considered represents a chance to prevent increases in mutual distrust (Lind and Tyler 1988). In Iitate Village after the earthquake disaster, the town had to deal with the dual problems of one-way communication and the loss of opportunities for dialogue.

Eliminating information asymmetries

Information asymmetries were also a major issue.[2] Provision of valid and reliable information is an essential requirement for public monitoring, a central function of civil society. If policies are advanced without providing equal access to information, then misunderstandings between residents and government will deepen, eventually resulting in the implementation of policies that diverge from public opinion and further fuel distrust; in the end, residents may give up on the locality completely. In an emergency situation such as this, some gaps in information between government and residents are to be expected. However, as noted above, even if disparities in information exist, they can be remedied, and it is possible to plan to avoid the breakdown in dialogue that occurred in Iitate.

Identifying power structures and political games

What has become clear through the above example is that the disaster-afflicted municipalities are unable to respond to the sentiments of their victimized residents in an acceptable manner. This inability did not derive simply from individual factors, such as a lack of opportunities for discussion between government and residents, or a misunderstanding of residents' needs based on a lack of capacities on the part of the government. As seen in the case of Iitate, the factors that inclined government toward both decontamination-centered policies of early-period reconstruction and the return of evacuated residents are deeply connected to power relations between the state and localities. These factors cannot be grasped without a deeper analysis that includes within its purview such factors as domestic energy issues, national economic policies, and national defense. At the very least, the subjects that are involved with reconstruction in Fukushima from here on in should be conscious of the ways in which civil society's capacity to perform its designated monitoring function is circumscribed by the above-noted structural factors.

Our responsibilities: public opinion as a collective assembly and its influence

Last, let us consider what we, as public citizens, need to do to facilitate reconstruction by noting a few challenging tasks ahead. First, it is necessary to objectively analyze the various problems affecting the daily lives of evacuees. Haruo Yamaura has problematized the fact that no one has organized or verbalized the current situation and problem structure of Fukushima in a form that is easily understandable (Fukushima Future Center for Regional Revitalization (FURE) 2014). Indeed, it is true that social-scientific analyses of the disaster have not yet clearly articulated the structure of damages outlined in this chapter.[3] In addition, policies regarding nuclear-accident damages and evacuees have relied almost entirely on natural scientific knowledge pertaining to radiation-exposure mitigation, while social-scientific analysis of problems and policies advanced in this chapter are still lacking (Yamashita *et al.* 2013). Accordingly, we must draw on interdisciplinary knowledge to further identify and clarify the structure of damages, and to analyze policy responses to the key tasks.

It is also necessary to adopt and support a multi-decade, long-term perspective on policy formation—resolution of the nuclear-power accident will take generations. Among the victims are more than a few voices asserting, for example, that while they will not return immediately, they would like to spend their retirement in their hometown, or that they would like to pass their hometown on to their children and grandchildren. However, in the discourse of current return and reconstruction policies, people with such feelings have been stigmatized as deserters or as people who are willing to throw their hometown away. Over the longer term, diverse patterns of return can be expected; these diverse choices should be recognized and institutional designs adapted to each scenario should be advanced.

In the effort to develop better means of implementing policies for rebuilding localities and lives in the future, the importance of external factors such as the political and legal system cannot be overstated. Nevertheless, as mentioned in the case of Iitate, current policies, systems, and forms of governance have put the afflicted areas in a highly disadvantageous position. In light of the fact that public opinion has played an influential role in shaping this situation, we cannot forget that the rift identified in this chapter, which has produced distrust and pushed some residents to give up, is facilitated by public actions and mindsets, including our insufficient understanding of the situation and aversion to political involvement. Indeed, it seems that is we who are at fault for the plight of the nuclear disaster-afflicted areas.

Acknowledgments

I would like to thank Masayuki Seto of Fukushima University for his assistance in completing this chapter. This work was supported by JSPS KAKENHI Grant Number 25220403.

Notes

1 The contents of this section are drawn mainly from Sato (2013, 67–85).

2 While criticism from residents has heightened, it cannot be said that officers and assemblymen have not attempted to respond. The local government has pursued a number of requests and negotiations with TEPCO and the national government, including quick responses to issues of compensation, redesignation of the evacuated areas in line with local conditions, and the establishment of timelines for return. However, as a result of the loss of opportunities for dialogue between residents and local governments, these efforts have not been visible to residents. Pushing forward without resolving information asymmetries—that is, pushing ahead with plans that are clear to government officials, but not understood by residents—is one of the situations that continues to spur distrust of the government.

3 The Science Council of Japan (2013, 5) points out that lack of verification of the actual state of evacuation in both the government and the Diet's investigations is one of the problems of policies adopted up to now.

References

Beck, Ulrick. 1997. "Subpolitics: Ecology and the Disintegration of Institutional Power." *Organization & Environment* 10(1): 52–65.

Dunlap, E. Riley. 1995. "Toward the Internationalization of Environmental Sociology: An Invitation to Japanese Scholars (Eyes on Environmental Sociology)." *Journal of Environmental Sociology* 1: 73–85.

Fukushima Future Center for Regional Revitalization (FURE). 2014. "Genpatsu Jiko Koiki Hinansha no Hatsugen wo Mochiita Shitsuteki Togoho (KJ ho) Bunseki Kekka Kara Mita Mondai to Kozo" [Identification of Problems and Structures of Affliction by the Application of the Qualitative Integrative Method (KJ Method) to Analyze the Remarks of the Evacuees of the Nuclear Accident]. *Fukushima Daigaku Utsukushima Fukushima Mirai Shien Center Nenpo* [Annual Report of the FURE]. Fukushima: Fukushima University. [In Japanese]

Funabashi, Harutoshi. 1999. "Kagai Katei no Tokushitsu: Kigyo Gyosei no Taio to Kagai no Rensateki Haseiteki Kaju" [Characteristics of the Victimizing and Suffering Process: Responses of Firms and Administration, and the Chain Reaction and Derivative Process of Victimizing]. In *Niigata Minamata Mondai: Kagai to Higai no Shakaigaku* [Niigata Minamata Disease Problem: Sociology of Victimizing and Suffering], edited by Nobuko Iijima and Harutoshi Funabashi, pp. 41–73. Tokyo: Toshindo. [In Japanese]

Habermas, Jurgen. 1985. *Lifeworld and System: A Critique of Functionalist Reason. Vol. 2 of The Theory of Communicative Action*. Translated by Thomas McCarthy. Boston: Beacon Press.

Hasegawa, Koichi, ed. 2004. *Constructing Civil Society in Japan: Voices of Environmental Movements*. Melbourne: Trans Pacific Press.

Iijima, Nobuko. 1984. *Kankyou mondai to higaisha undou* [Environmental Problems and Sufferers' Movements]. Tokyo: Gakubun-sha. [In Japanese]

Iitate Village. 2016. Iitate Mura: Shinsai Iko no Iitate Mura wo Tsutaeru Joho Saito [Information Site to Communicate the Situation of Iitate Village after the Disaster]. Accessed January 25, 2016. www.vill.iitate.fukushima.jp/saigai/wpcontent/uploads/2015/12/39d4302b0551b46607506ad71c3eeaa9.pdf. [In Japanese]

Imai, Akira. 2011a. "The Primary Survey of Inhabitants Who Were Evacuated from the Nuclear Power Plant Disaster." *Monthly Review of Local Government* 393: 1–37. [In Japanese]

Imai, Akira. 2011b. "The Second Primary Survey of Inhabitants Who Were Evacuated from the Nuclear Power Plant Disaster." *Monthly Review of Local Government* 398: 17–41. [In Japanese]

Imai, Akira. 2012. "The Third Primary Survey of Inhabitants Who Were Evacuated from the Nuclear Power Plant Disaster." *Monthly Review of Local Government* 402: 24–56. [In Japanese]

Kanno, Masashi, and Ryusuke Takaki. 2012. "Higashi Nihon Daishinsai ni okeru Naraha-machi no Saigai Taiou (1): Community no Saisei ni Mukete" [Disaster Response of Naraha Town to the Great Eastern Japan Earthquake Disaster (1): Toward the Regeneration of Communities]. *Research Bulletin of Iwaki Meisei University. Graduate School of Humanities* 10: 36–51. [In Japanese]

Kawazoe, Saori, and Masaki Urano. 2011. "Impact of the Nuclear Power Plant Disaster and Issues for Reconstruction: A Case Study of Iwaki City with Diverse Communities." *Nihon Toshi Gakkai Nenpo* [Annals of the Japan Society for Urbanology] 45: 150–159. [In Japanese]

Lind, Allan E., and Tom R. Tyler. 1988. *The Social Psychology of Procedural Justice*. New York: Plenum Press.

Matsui, Katsuhiro. 2012. *Shinsai Fukko no Shakaigaku -Futatsu no "Chuetsu" kara "Higashi Nihon" he* [Sociology of Disaster and Reconstruction: From the two "Chuetsu" to "Higashi Nihon"]. Tokyo: Liberta shuppan. [In Japanese]

Ohashi, Yasuaki, and Ryusuke Takaki. 2012. Higashi Nihon Daishinsai ni okeru Naraha-machi no Saigai Taiou (3)—Kyouiku Kinou no Iji Saihen [Disaster Response of Naraha Town to the Great Eastern Japan Earthquake Disaster (3): Retention and Restructuring of Educational Functions]. *Research Bulletin of Iwaki Meisei University. Graduate School of Humanities* 10: 63–74. [In Japanese]

Sato, Akihiko. 2012. "Zenson Hinan wo Megutte—Iitatemura no Kunou to Sentaku" [On the Whole-Village Evacuation: Struggles and Choices in Iitate Village]. In *Genpatsu-hinan Ron: Hinan no Jitsuzou kara Sekando Taun, Furusato Saisei Made* [Study of Nuclear Evacuation: From the Realities of Evacuation to Second Towns and the Restoration of Home Towns], edited by Yusuke Yamashita and Hiroshi Kainuma, pp. 91–137. Tokyo: Akashi Shoten. [In Japanese]

Sato, Akihiko. 2013. "Refugees from Radioactivity, the Return to Iitate, and Conflicts between the Village and the Refugees over the Recovery from the Disaster." *Social Policy and Labor Studies* 4 (3): 38–50. [In Japanese]

Science Council of Japan. 2013. "Genpatsu Saigai kara no Kaifuku to Fukko no Tameni Hitsuyou na Kadai to Torikumitaisei ni Tsuiteno Teigen" [Recommendations for the Necessary Measures toward the Recovery and Reconstruction from the Nuclear Disaster]. Accessed June 27, 2013. www.scj.go.jp/ja/info/kohyo/pdf/kohyo-22-t174-1.pdf. [In Japanese]

Shinohara, Hajime. 2004. *Shimin no Seijigaku—Tougi Demokurashii towa Nanika* [A Study of Citizen Politics: What is Deliberative Democracy?]. Tokyo: Iwanamishoten. [In Japanese]

Sorensen, André, and Caroline Funck, eds. 2007. *Living Cities in Japan: Citizens' Movements, Machizukuri and Local Environments*. Oxford: Taylor & Francis.

Takaki, Ryusuke, Akihiko Sato, and Masayoshi Kato. 2012. "Fukushima Daiichi Genpatsu Saigai no Genjou to Fukko Kadai" [Current Status of the Fukushima Daiichi Nuclear Power Plant Disaster and Problems for Reconstruction]. Paper presented at the Shakaigaku 4 Gakkai goudou Sinpojiumu: Nihon Gakujutukaigi Raundo Teiburu Houkoku Shiryo [Symposium of the Four Social Science Disciplines: Reports for the Science Council of Japan], Tohoku University, July, 2012. [In Japanese]

Tamba, Fuminori. 2012. "Study of Eight Towns and Villages in Futaba County Reveals Situation of Evacuees due to Fukushima Daiichi Accident." *Kankyo to Kougai* [Research on Environmental Disruption] 41 (4): 39–45.

Tomioka Children's Future Network. 2013. *Tomioka Kodomo Mirai Nettowaaku Katsudou Kiroku*. [Records of Activities of Tomioka Childrens' Future Network]. Unpublished Report. [In Japanese]

Yamashita, Yusuke, and Hiroshi Kainuma, eds. 2012. *Genpatsu-Hinan Ron Hinan no Jitsuzou kara Sekando Taun, Furusato Saisei Made* [Study of Nuclear Evacuation: From the Realities of Evacuation to Second Towns and the Restoration of Home Towns]. Tokyo: Akashi Shoten. [In Japanese]

Yamashita, Yusuke, Ichimura Takashi, and Akihiko, Sato. 2013. *Ningen Naki Fukko*. [Reconstruction without Human]. Tokyo: Akashi Shoten. [In Japanese]

Yanagisawa, Takashu, and Kikuchi Mayumi. 2012. "Higashi Nihon Daishinsai ni okeru Naraha-machi no Saigai Taiou (2)—Hinansaki ni okeru Fukushikinou no Iji to Kazokukinou no Saihen ni Mukete" [Disaster Response of Naraha Town to the Great Eastern Japan Earthquake Disaster (2): Toward the Retention of Welfare Functions and the Restructuring of Family Functions at Evacuation Destinations]. *Research Bulletin of Iwaki Meisei University. Graduate School of Humanities* 10: 52–62. [In Japanese]

Yokemoto, Masafumi. 2013. *Genpatsu Baisho wo Tou—Aimai na Sekinin, Honrousareru Hinan-sha*. [Questioning Nuclear Accident Compensations: Ambiguous Responsibilities and Trifled Evacuees]. Tokyo: Iwanami-shoten. [In Japanese]

4 Why do local residents continue to use potentially contaminated stream water after the nuclear accident?

A case study of Kawauchi Village, Fukushima

Takehito Noda

Crisis of village communities after the nuclear plant accident

This chapter investigates why the residents in a village that was affected by the Fukushima nuclear accident insist on continuing to use stream water from the mountains even though the water may be radiation-contaminated. I focus on the *Yamanokami* ("spirits of the mountain") Water Supply System Association (YWSSA), a community organization in Kawauchi Village, Fukushima, which is located within the 20 to 30 km radius zone of the TEPCO Fukushima Daiichi Nuclear Power Plant (NPP). Kawauchi Village was designated as an evacuation area after the hydrogen explosions at the Fukushima Daiichi plant resulting from the massive earthquake and tsunami on March 11, 2011. The entire village was evacuated by March 16, and the residents were forced to scatter across temporary shelters for about ten months. By January 2012 the village office had restored administrative services and issued a call for residents to return to the village. Nevertheless, as of March 2015, four years after the nuclear accident, only half of the former villagers have returned home.

The village community now faces a fateful crisis, one that does not merely include the state of diaspora in which many of the former villagers are living. Rather, it is the kind of crisis best described by Naomi Klein, who coined the term "disaster capitalism" to describe how commercial capital uses the recovery process from such disruptive events as terrorist attacks, war, and natural disasters as a prime to encroach upon the public sphere (Klein 2007). In particular, Klein argues, neoliberal economists and political leaders influenced by the Chicago school of economics use post-disaster crises to push through radical social and economic reforms based on the principles of libertarian free-market economics. In Japan, post-disaster policies are often framed as efforts to "rationalize" (*gorika*) existing social and economic institutions and practices.

While Klein focuses primarily on market-oriented (rationalization) policies introduced by agents outside of the communities in crisis, rationalization may be also pushed forward within the communities. The environmental sociologist Kiyoshi Kanebishi analyzes village communities after the Great East Japan

Earthquake Disaster and makes a case for the presence of the "internal shock doctrine" (Kanebishi 2014), which disrupts and in some cases destroys the organizing principles of community members' daily lives. Accordingly, the crisis faced by these disaster-afflicted communities refers to the situation in which various community-based "life organizations" are restructured or dissolved in order to adapt them to opportunistic rationalization policies that exploit the post-disaster chaos. As the result, the very existence of the communities comes under threat.

Seikatsu soshiki ("life organization") is an analytical concept coined by Kizaemon Aruga (1968), who laid the foundations of modern Japanese sociology.[1] Life organization is a type of social organization that is intricately related to the daily lives of a local community's inhabitants. Aruga distinguishes *life organization* from *social organization* by emphasizing the former's focus on the role of shared values among members of an organization. He calls a set of these values *seikatsu ishiki* ("life consciousness"), which is essential in the study of lives of local communities' inhabitants because it serves as a normative standard to sift through externally induced changes brought upon them (e.g. increased pressure to embrace more market-based institutions).

The Sawa hamlet of Kawauchi Village, the community on which this study focuses, has come under pressure to rationalize its life organizations as part of post-disaster recovery efforts. After the evacuation and scattering of the residents, its *jichikai* (neighborhood association) has become non-functioning, and other life organizations such as its ritual festival organization and *soshiki-gumi* (funeral associations) have been practically dissolved. Despite these difficult circumstances, this community has made an exception for its local water-supply system association, and is trying to keep it in operation. Given that the source of this water supply is within 20 km of the Fukushima Daiichi plant, and that the residents have been advised by the village office not to use this water due to possible radiation contamination, their resistance to abandoning the traditional water system is both puzzling and worthy of investigation.

This chapter therefore aims to articulate the local communal logic of sustaining the water-supply system association, as a life organization, through an in-depth analysis of livelihoods in a local community. Following Aruga's theoretical framework, this chapter focuses on the normative values of those who support the continuation of the water-supply system association as a way of articulating a communal logic to cope with the effects of the nuclear disaster. This study is a qualitative analysis of information gained through interviews with residents of Kawauchi Village, conducted intermittently over the time span of December 2013 through June 2015.

Local residents' reactions to the nuclear accident

Overview of Kawauchi Village

Kawauchi Village occupies a basin located in the central plateau of the Abukuma Mountains of Fukushima. Approximately 90 per cent of the area is forested and only 5 per cent is farmland. The village population is 2,794 people as of October

2013. The village has eight municipal districts and 24 hamlet communities.[2] The Sawa community, which maintains the local water-supply system, is a small hamlet of 21 households centrally located in the third district of the village. The village economy was once centered on forestry, supplemented by crop and dairy farming. During the agricultural off-season, villagers worked away from home in cities such as Tokyo. In the mid-1960s, construction of the Fukushima Daiichi and Daini Nuclear Power Plants began in the neighboring towns along the coast. Since then a large number of villagers have been employed by the nuclear power plants and related businesses. For good or bad, villagers say, they have been living under the shadow of the nuclear power plants. Yet few could ever foresee or imagine such a disastrous accident as that which took place at the Daiichi plant.

The Great East Japan Earthquake and post-accident migration of people

Table 4.1 shows a timeline of events involving Kawauchi Village and its residents immediately after the earthquake. The earthquake occurred on March 11, 2011 and was reportedly a 6-lower on the Japan Meteorological Agency seismic-intensity scale in the village. Located inland and sitting on relatively firm ground, Kawauchi Village suffered less damage than most of the surrounding areas. The villagers testify, nevertheless, that the shaking they experienced was unprecedentedly terrifying.

On the following day, March 12, 8,000 residents of the coastal town of Tomioka, located near the damaged Daiichi plant, were evacuated to Kawauchi, just to its west. The villagers were pressed to take care of evacuees from Tomioka. The women of the Sawa community volunteered to cookout at one of the local shelters. They also brought food from home, and offered blankets and futons for the evacuees. At 3:36 p.m. on that day, Unit 1 of the Daiichi NPP exploded, but most of the Kawauchi villagers did not learn about the hydrogen explosion immediately as they were busy helping the evacuees. However, even after having learned about the accident through media later in the day, they did not feel a sense of extreme urgency, since “there was little that could be done at that point” (woman in her seventies).

On March 14, another hydrogen explosion occurred at Unit 3 of the Daiichi plant, and people began to worry. By this time some villagers and Tomioka evacuees had started to leave the village voluntarily. On the morning of March 15, Unit 4 experienced a hydrogen explosion and the government issued an indoor evacuation order, which directed all residents within 20–30 km of the Daiichi plant to stay indoors. Because all of Kawauchi Village lies within this zone, the villagers could no longer provide relief assistance to Tomioka evacuees. By mid-afternoon, the village office had strongly advised all residents to evacuate on their own. Those who could evacuated the village soon after, typically to stay with their friends and families living elsewhere. On March 16 the village mayor decided to order the entire village to evacuate, and the remaining villagers as well as Tomioka evacuees relocated themselves to Koriyama City, about 40 km away. Interestingly, when the village office issued an evacuation order, no hamlet community-level decisions were made or collective measures taken (which is rather unusual) and the village office left evacuation-related decisions up to each individual household. This began the diaspora of the villagers from March 16 onward.

Table 4.2 shows major events leading up to April 2012, when the village office resumed its functions and issued a call for residents to return to the village after approximately a year of life in evacuation.

Kawauchi Village was the first among the municipalities affected by the nuclear accident to call for its residents to return to formally evacuated areas. In addition to the lowered risk of additional severe nuclear accidents, the level of

Table 4.1 Timeline of events pertaining to Kawauchi Village immediately after the earthquake

Year	*Date*	*Time*	*Events*
2011	March 11	14:46	Tohoku earthquake occurred (lower-6 measured in Kawauchi)
	March 12	5:44	Evacuation order issued for 10 km of Fukushima first plant (3 km within second plant)
			8,000 Tomioka residents evacuated to Kawauchi
		15:36	Fukushima Daiichi plant unit 1, hydrogen explosion
		18:25	Evacuation order for 20 km radius zone
	March 14	11:01	Fukushima Daiichi plant unit 3, hydrogen explosion
	March 15	6:10	Fukushima Daiichi plant unit 4, hydrogen explosion
		11:00	Kawauchi designated as restricted and evacuation zone
		15:00	Voluntary evacuation advised to Kawauchi villagers
	March 16	Early morning	Kawauchi and Tomioka group evacuation to Koriyama

Source: Kawauchi Village Decontamination and dose control for repatriation from the nuclear power plant accident (October 1, 2014).

Table 4.2 Events leading up to the call for return

Year	*Date*	*Events*
2011	September 30	Order for emergency evacuation preparation zone is lifted (within areas of 20–30 km radius of the Daiichi plant)
		Creating recovery planning, reconstruction vision, decontamination planning
	October	Conference with residents
	November	Start of decontamination
2012	January	Declaration of repatriation
	March	Village administration reopens
	April	Return of villagers
		Schools and medical facilities including clinic centers reopen, and new bus lines are opened

Source: Kawauchi Village Decontamination and dose control for repatriation from the nuclear power plant accident (October 1, 2014).

radiation contamination in the village had been relatively low compared to that in neighboring municipalities (because of the wind direction when large volumes of radioactive particles were released). Nevertheless, the number of returnees remains lower than the village office had hoped. By October 1, 2013, the number of returnees had reached 1,455, which represents 52.1 per cent of all former residents. That is, it was two and a half years before more than half of the former residents returned. But, if we limit the figure to those who completely left their temporary residence—excluding the "dual-residence holders" who still maintain another residence (e.g. shelters and rental properties) elsewhere—the number of returnees was only 535. Today, four years after the accident, it is estimated that only about half of the former residents have moved back to the village.

Let us look at the Sawa hamlet community more specifically. Of the 24 households, about half have returned completely. Others still live in temporary housing in other parts of the village or in cities such as Koriyama and Iwaki, and commute to their houses a few days a month. In other words, there are a sizable number of dual-residence holders.

It is noteworthy nevertheless that some Sawa community members had already voluntarily returned to their houses before the official call for return. Some came back as early as ten days after the accident. Such an early return was possible, according to some villagers, because they were able to secure access to drinking water in the village. Kawauchi Village is one of the few municipalities in Japan that have not established centrally controlled municipal water systems.[3] The villagers have always obtained their drinking water from wells and stream water from mountains through simpler and more localized water-supply systems. Unlike residents of many other municipalities whose water systems were damaged by the earthquake, people in Kawauchi Village were able to secure drinking water at the hamlet scale, making it possible to continue living in the village even after the accident. The sociologist Atsushi Sakurai (2014) studied an afflicted community in Iwate Prefecture after the Great East Japan Earthquake Disaster, and concluded that the community's survival strategy was intricately related to the availability of local water sources such as well and stream water from mountains. It took only two days for the community that he studied to build a localized water system by using the surface water available, while it took two and a half months to restore its municipal water system. The localized water source enabled the local residents to sustain their livelihood with little interruption after the disaster. Indeed, the availability of such local water sources, rather than having to depend on public water works, was one of the key reasons that the Kawauchi village office decided to call for the repatriation of residents, according to a village office worker in Kawauchi.

Life with the *Yamanokami* water

People in Kawauchi Village have traditionally used stream water from the mountains for all aspects of their livelihood. A study group led by Masuo Ishikawa says that people also started digging wells to obtain water for domestic usage relatively recently, in the late 1920s (Ishikawa, Kozako, and Suzuki 2004).

In the course of modernization, the main water source shifted from stream water to underground water. Yet there exists one group of people in the Sawa hamlet of the village who continue to use stream water from the mountains: the *Yamanokami* Water Supply System Association, which is now the only remaining association of this kind in the village. The Sawa community worships the spirits of the mountains (*yama no kami*) from which the stream water comes, and the association was named after the spirits, who, it was hoped, would protect the water forever. Such faith in the spirituality of nature is waning today. A resident in his seventies who takes care of the *Yamanokami* shrine says that his parents' generation used to have ceremonial festivals, which included sacred dance and music, but these activities no longer take place due to the aging of the guardians and a lack of interest among younger people.

The *Yamanokami* Water Supply System Association was established in 1963. Figure 4.1 depicts the location of the water system relative to the hamlet community. Originally there were 20 member households in the association, and the water was supplied to each house. At that time these members included some households outside of the Sawa community that did not have their own wells or other water sources. These outside-community members eventually left the association (presumably as they secured their own community water sources). Currently the association has 15 member households, including ten households in Sawa and five in two neighboring communities, and they continue to use

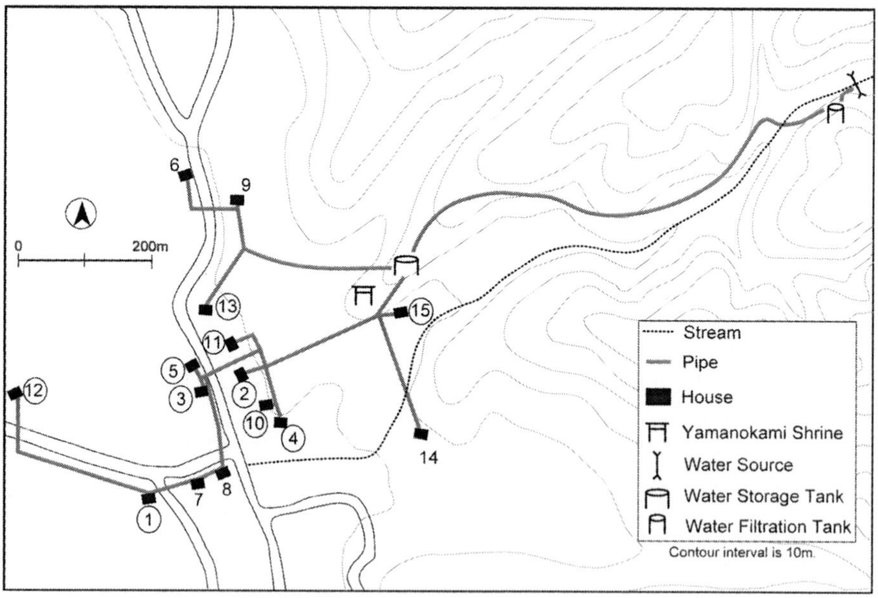

Figure 4.1 Sawa community and *Yamanokami* water system.

Note: Figure shows the locations of all 15 Yamanokami member households. Numbers indicating the ten households of the Sawa community are circled.

this water from the mountains almost half a century since the inception of the system. Figure 4.1 shows the locations of all 15 *Yamanokami* member households, among which ten households of the Sawa community are circled.

The water-supply system was designed and built entirely through the collective labor of the association members. The project took two months to complete. The system makes good use of the terrain and the slope of the mountain. Water from a small creek in the mountain is first channeled into a large concrete filtration tank about 1 km away from the hamlet. The bottom of the tank is filled with charcoal, pebbles, and gravel for filtration. (One member still remembers how hard it was to hike up the mountain with a large wooden box full of gravel to make concrete for this filtration tank.) Filtered water then runs through an underground pipe along the side of the mountain to the water-storage tank just behind the *Yamanokami* shrine, from which two distribution pipes supply water to each household. The system runs by water pressure only, without any additional sources of power; hence several air-vents and in-water debris removal screens, without which stalled water flow may result, are installed along the pipe. It is a well-designed, sophisticated water-supply system. Following its installation, the association set up regular maintenance rules in which a group of three members, in rotation, cleans up the water source and tanks every month, and all member households participate in an annual cleanup day in April. It became a quaint, small tradition of the community to have a gathering under cherry-blossom trees after the annual cleanup. Elderly members recall that this used to be one of the few delightful activities in the community to which everyone looked forward.

Table 4.3 Status of *Yamanokami* water usage and evacuee return (August, 2014)

	Yamanokami water uses	*Private well*	*Status*
1	Domestic•Field	Y	Full return
2	Domestic•Field	Y	Dual residence
3	Domestic•Field	Y	Full return
4	Domestic•Field	Y	Full return
5	Domestic•Field	Y	Dual residence
6	Domestic•Field	Y	Full return
7	Drinking•Domestic•Field	N	Dual residence
8	Domestic•Field	Y	Dual residence
9	Drinking•Domestic•Field	N	Dual residence
10	Domestic•Field	Y	Full return
11*	(Drinking)•Domestic•Field	Y	Full return
12	Domestic•Field	Y	Full return
13	Domestic•Field	Y	Dual residence
14	Domestic•Field	Y	Dual residence
15	Domestic•Field	Y	Dual residence

Source: Based on interviews with villagers (author).

Note: The elderly couple at Household 11 has been drinking the water after the disaster, but has recently installed a well in preparation for their son's family's anticipated return.

Table 4.3 lists the use of *Yamanokami* water and the evacuation/return status by household.[4] Of the 15 association-member households, seven "fully" returned to the Sawa hamlet, which means that these households completely moved out of their temporary housing elsewhere. All member households continue to use the *Yamanokami* water for at least some purposes, such as domestic use and farming. It is particularly noteworthy that a few households have continued to drink the water even after the nuclear accident. Just recently, one household (Household 11 in Table 4.3) installed a private well. This is because the village office now advises residents not to drink stream water due to concerns regarding possible radioactive contamination, and offers a subsidy of up to 1 million yen for the installation of a new private well. Despite such an administrative push, a few residents of Sawa continue to drink the *Yamanokami* water, and even those who have a private well insist that it is important to sustain the use of the *Yamanokami* water in some ways. Accordingly, although many life organizations of the community have been "rationalized" and even dissolved, the community members seem committed to upholding the association.

Significance of the *Yamanokami* Water Supply System Association

To understand the association members' insistence on using the water, I focus on three issues: the role of the *Yamanokami* association as a critical life organization in the everyday life of the community; changes in the organizational rules of the association after the disaster; and narratives of the local residents who continue to use the *Yamanokami* water against the advice of the local office.

The role of the Yamanokami Association as a life organization

As mentioned above, only about half of the former residents have returned to the village. Consequently, many of the village communities' life organizations are not functioning properly. Under these circumstances, various agencies are at work—both internal and external to the village—to restructure these life organizations so as to help restore and ease the lives of the residents who returned. There are four life organizations that are particularly central to the lives of the residents in Kawauchi Village: municipal districts (lower administrative units of the village); hamlet communities within these districts; funeral associations; and water-supply system associations (although the last organization remains only in the Sawa hamlet today). Below I compare the role of the *Yamanokami* Water Supply System Association and its changes after the disaster to those of other life organizations in the Sawa hamlet community.

The Sawa hamlet, made up of 15 households—of which ten belong to the *Yamanokami* association—is located in the third municipal district of the village. The scale of a hamlet makes it the most familiar life organization for residents as the one within which most everyday interpersonal interactions traditionally occur. Nevertheless, with the aging and shrinking population of the hamlet, the hamlet community's capacity for self-government was weakening even prior to the disaster. This hamlet used to organize various community activities such as a New Year party and group trips, but those are things of the past. It used to pass along a community bulletin board in a circle from household to household, but this practice disappeared with the implementation of the wireless disaster-warning system some time ago. Nowadays residents may seldom see each other even in their own neighborhood.

If the hamlet has been losing its importance to residents in recent years, the municipal district has long lost its significance as a life organization for many residents of the village. The municipal district now acts simply as an administrative channel of the village office; it rarely works as an arena in which various local problems are discussed and solved, and in which residents get acquainted and interact with each other.

It is not surprising, then, that both the hamlet and the municipal district became completely nonoperational after the earthquake and the nuclear accident. This is evidenced by the complete absence of initiatives at these levels to discuss and strategize for post-disaster recovery, and by the fact that no district fees have been collected since the disaster. The disaster was the final blow to these life organizations, which were already weakening and losing relevance.

Two occasions that continued, although just barely, to function as media of socialization among local residents were funerals and cleaning of the water-supply system. Burial was being still practiced in Kawauchi Village only until about 20 years ago (despite the government's push toward cremation), and funerals continue to be community-managed events. In Japan, funerals have been traditionally managed by funeral associations, which are reciprocal organizations made up of neighboring residents. The membership and areal extent of *soshiki-gumi* do not necessarily coincide with those of hamlets and water-supply associations. Each of these life organizations have slightly different membership compositions. The *soshiki-gumi* in the Sawa hamlet used to collect association fees of about 1,000 yen per month until it was equipped with all necessary supplies for funerals, such as pole tents, tables, an outdoor gas stove, cooking utensils, and tableware. Nevertheless, during the past 20 years an increasing number of households have begun to have their funerals at funeral homes in or outside of the village, rather than at their own houses. The disaster and nuclear accident prompted the members of the *soshiki-gumi* to rationalize (i.e. dissolve) the organization. At its dissolution, the *soshiki-gumi* members bought up all the supplies, using the money to pay off the rents for the storage, and the remaining money was allocated equally among the members.

The life organizations discussed above were non-functioning and practically dissolved by the time the village was hit by the disaster and the nuclear accident. In contrast, the *Yamanokami* association is a still-functioning organization and played an important role in the residents' lives. Prior to the disaster, the monthly cleanup by a team of three members and the all-out annual cleanup and weeding were important opportunities for the residents to communicate with each other. Indeed, as exemplified by the fact that the annual cleanup was the place where the dissolution of *soshiki-gumi* was discussed after the disaster, the water association has been offering a space in which residents can talk about assorted problems and needs of the community. At monthly cleanings, too, members talk about—among various topics of gossip—other members' health and villagers' whereabouts. In addition, the association sends a circular among members in order to notify the next monthly cleaning team that it is their turn. Accordingly, the Yamanokami association has continued to be a vital life organization for the residents even after the disaster.

Why is it that the water-supply system association, and not other life organizations, has endured till now? It is because this organization comprehensively supports both spheres of production and living through the use of surface water from the mountains, which the residents continue to need even after the nuclear accident. Nevertheless, it has been far from easy to manage the water because not all the association members, who were once all forced to evacuate, came back home after the call for return. Routine cleanups normally take place on weekends and holidays. But one can easily imagine that those who live apart from their family members in other towns and cities—as it is often the case that younger people with children typically prefer to live farther from Fukushima Daiichi—would rather spend their weekends and holidays with their families. Even for those who would normally put community activities before their own needs, private issues have increased in importance since the disaster. The association began to make adjustments to its organization in order to sustain the management of the water.

Efforts to sustain the life organization

After the nuclear accident, the *Yamanokami* association made three important decisions. First, it cut back the regular cleanup from once a month to once every other month. Second, it decided to exempt women who live alone from the routine obligation to be a board member. Third, it reaffirmed that it would continue the rotation system for board members.

These decisions are evidenced by the record of the association's activities after the disaster. Table 4.4 shows the rotation list of cleanup duties between January 2011 and March 2014. From this list, it appears that the association was idle for about a year and four months after the disaster, until July 21, 2012. Yet, as already mentioned above, some residents were already returning to the village even before the village office's call for return. Indeed, only ten days after the evacuation, one of the association's leaders returned to his home with his wife. Their son, who used to live with them, remained in evacuation.

This household has relied for all of its water needs, from domestic use to drinking, on the water from the mountains. Even if there may be concerns about possible radioactive contamination, the water was essential for them to live in the village. Hence, although the association temporarily stopped working, this couple occasionally checked and removed debris from the water tank to maintain the system. Their house also became a meeting place for those who temporarily visited their homes from their evacuation sites. This meant that the couple's house became a hub of information to do with issues such as who was evacuating where and who was coming back at what time, encouraging more residents of the Sawa hamlet to return permanently.

Table 4.4 *Yamanokami* water-supply association: duty roster (January 2011–March 2014)

Date		*Memo*
2011	01–	Leaves removed from water intake. Satisfactory.
	02/27	Satisfactory.
		The Great East Japan Earthquake; nuclear power plant accident occurred
2012	07/21	All members attended to clean (after 1 year and 4 months) and mowed the lawn. Satisfactory.
	08/26	Leaves and mud removed from water intake. Each tank cleaned. Satisfactory.
	09/28	Leaves and mud removed from water intake. First and second tanks cleaned. Water level adequate. Satisfactory.
	10/27	Drain outlets in upper and lower tanks cleaned. Satisfactory.
	11/30	Water intake of each tank cleaned. Satisfactory.
	12/30	Water intake of each tank cleaned. Satisfactory.
2013	02/23	Water intake of each tank cleaned. Satisfactory.
	03/28	Water level low. Leaves removed. Mud removed OK in each tank. 2 items in second tank, satisfactory.
	04/28	All tanks cleaned. Satisfactory.
	05/26	Tanks cleaned. Satisfactory.
	06/23	Water intake, each tank cleaned. Satisfactory. Water level low.
		Memo describing "bi-monthly from August" schedule
	09/25	Leaves removed. Tanks cleaned. Satisfactory.
	10/25	First and second tanks checked, no issues found. Water level high due to heavy rain.
	11/27	Washing leaves and mud removed from first tank water intake. Mud removed in second tank. Water level adequate. Satisfactory.
2014	01/25	Leaves removed at entrance. First tank drained 5 times. Second tank, mud removed.
	03/30	Water intake and tank checked. Satisfactory.

Source: Based on interviews with Yamanokami members (author).

On July 21 the *Yamanokami* association resumed cleanup work as an association activity, and also cleared the path to the water source for the first time since the disaster. On that day members decided, after some discussion, to resume normal management of the water-supply system. Initially the cleanup was undertaken on a monthly basis, but it was reduced to a bi-monthly basis starting in August, 2012 (Table 4.4) because about half of the members who had come from their temporary housing outside of the village to clean up could not spare the time. It is important to note that the most important reason for the association's first decision to lower the frequency of cleanups was to avoid the loss of its members.

On April 27, 2014, all member households participated in the all-out cleanup for the first time since the disaster. Remarkably, every household, including those that were still living away from the community, took part. After the cleanup, a general meeting was held and it was decided to exempt single-living women from board-member (i.e. chair) duties. Normally the term of the association chair is two years, and it rotates among all member households. His/her tasks range widely, from acting as a liaison and coordinator to doing all other miscellaneous work for the association. The decision to exempt single-living women from this responsibility was a measure to reduce a burden for the socially vulnerable. This second decision was also an effort on the part of the association to sustain its functioning without any "dropouts."

The third decision was to reconfirm continuation of the system of rotating board members. This was a response to a suggestion made by one member at a meeting after the disaster. He proposed that, instead of continuing the rotation system, the acting caretaker of *Yamanokami*, who was one of the most central figures of the association, could become the permanent chair, given the difficult and diverse range of circumstances that many members were facing. In other words, this was a proposition to rationalize the organization. Some members were initially supportive of this proposition, but the acting caretaker was against it. Why? It was not that he wanted to avoid the responsibilities; rather, it was because he knew that the association would dissolve if all members did not equally share responsibilities through the rotating board-member system. That is, the rotation system was seen as a necessary rule to sustain the organization. If only particular members took up the role of board member, others without such responsibilities would be more inclined to leave the association. The members eventually came to agree on the importance of the current system, which acted as a system of shared responsibility based on the assumed equality among the members. By the end of the meeting, the members had unanimously agreed to continue it.

The *Yamanokami* Water Supply System Association made these collective decisions in order to sustain the organization while mitigating burdens on its members. These decisions to amend some organizational rules and reaffirm another have shared characteristics; they all imply the substantive equality of the constituent members. Such equality is a well-known characteristic of rules of resource use in Japanese rural societies (Ikegami 2007). The decision to exempt single-household women from board membership took into consideration their particular disadvantages resulting from the disaster. The purpose of such measures is to equalize the burden of work borne by members in order to sustain the organization.

Determination not to dig a new well

It is not only elderly association members who value organizational sustainability backed by substantive equality in labor. Younger members are also sympathetic to such logic, and the sentiment is symbolically captured by the assertion "we must continue to drink this water." The youngest member of the association, a 40-year-old male, is determined not to build a new well. After the nuclear accident, the village decided to offer subsidies of up to 1 million yen for any villagers who wanted to dig a private well as a response to possible contamination of surface water. Accordingly, the man was advised by the village office to build a well using the subsidies, but he refused and continues to refuse to build a new well. Why does he insist on drinking potentially radiation-contaminated water?

He used to live in a three-generation household, but the family members scattered after the disaster. Immediately after the disaster, he, his near full-term pregnant wife, and their child evacuated to Tokyo, where his sister lived. His mother evacuated to Saitama Prefecture, but returned to the village about a year later because she could not get used to the change in environment. His aged grandmother lives in a home for the elderly in a neighboring town. He returned to Kawauchi by himself one month after the disaster because he ran his own electric-contracting business. After his wife gave birth to a new baby, he moved with his wife and children to temporary housing in Iwaki City, from which he has been commuting to Kawauchi on a daily basis.

This means that he has been drinking the water from the mountains over the past four years. His friends worry about him drinking the water from mountains within the 20 km radius of the exploded nuclear plant; they voiced so much concern that he once tested the water at the source, but no radioactive substance was detected. At another time he cooperated with the radioactivity-detection test conducted by Nagasaki University and had the water tested using a filter attached to the faucet in his house. A small amount of radioactive substance was detected, but the level of radioactivity did not exceed the safety limit set by the government. Since then he has been declining offers and recommendations for additional tests from various sources. The reason that no radioactive substance was detected in the water at the source is because the radioactive substance itself does not dissolve in water. However, because it adheres to other objects, it could likely be detected in sand and mud at the bottom of the creek. This man is also aware that, although no radioactive substance had been detected in water in the storage tank, it could be detected, probably in an amount that does not exceed the government-set standards, in the sand at the bottom of the tank. Yet, he says that he has no further plans to test it. Moreover, he believes that this would be a bad idea, because if a radioactive substance is ever detected—no matter how small the amount is—it will create a reason to avoid drinking this water. As a result, some people may stop drinking or using the *Yamanokami* water and some may leave the association, which would make the association more difficult to sustain. This is precisely why the man is determined to keep drinking the *Yamanokami* water, even if he becomes the last person to do so.

What this story makes clear is that the reason that the man continues to drink the *Yamanokami* water is not that he values the water itself as a natural resource; rather, it is that he values the very presence of those who have long maintained the *Yamanokami* water-supply system. But, then, why is it so critical to sustain the association? The man goes on to say that he came to feel even more strongly about the need for the association after his experience of the disaster. He became painfully aware that real support in times of crisis comes from the people in local communities and not from administrative branches of governments, such as the state, prefecture, or village. For example, when he evacuated he could not take his dog, and the only people he could count on to feed his dog were his neighbors who remained in the village. This seems like a trivial example, but precisely because such things are a part of mundane everyday life, he feels it necessary to maintain the interpersonal relationships on which he can depend.

As described earlier, many life organizations in the hamlet have become completely dysfunctional since the disaster. The man discussed was also already keenly aware that the local society was eroding, and that is why he thinks that the Water Supply System Association should become a critical life organization in the process of rebuilding the community. He also understands his positionality and responsibility. Most *Yamanokami* members are elderly people. Seeing how they speak of the need to sustain the association and continue to put their labor into the association's work, he feels strongly about the responsibility of the younger generation. He wants to be helpful to other members in need, and feels obligated to look after them. Because of the nature of his business, he could work outside Kawauchi; yet, he chooses to commute to Kawauchi because of the Water Supply System Association. In short, his decision to drink *Yamanokami* water reflects his determination to uphold his relationships with the community residents.

Conclusions

The purpose of this chapter has been to explain why those who have suffered from the nuclear disaster continue to use potentially radiation-contaminated surface water by focusing on the logic of sustaining the local water-supply system association. Major findings are summarized as follows.

The Sawa hamlet, the case-study site, became an arena of rationalization policies, which originated both externally and internally, under the name of rebuilding after the disaster. In this crisis, various life organizations dissolved. However, the *Yamanokami* Water Supply System Association avoided such a crisis because it was able to function as a life organization that embraces all the residents' aspects of living. That is, the Water Supply System Association has historically played an integral role in the residents' production and reproduction (i.e. living) through the use and management of the water. Moreover, it is an organization that thoroughly adheres to the principle of substantive equality among the members. The modifications of the management rules after the disaster well illustrate this point. Using the water association as a basis, the residents are striving to rebuild their lives and livelihoods in the village. This is why they insist on using the potentially contaminated water from the mountain; its continued use means the maintenance of interpersonal relationships within the association.

More than four years have passed since the earthquake and nuclear accident. Kawauchi Village is revisiting its past policy guidelines and is contemplating the installation of a public water system as part of its post-disaster recovery and long-range planning (The 4th Kawauchi Village Comprehensive Plan, March 2014). However, as this chapter has shown, the Water Supply System Association is not simply an organization to supply water for the residents; rather, it has begun to function as a core life organization to support the rebuilding of a post-disaster community life. As local communities face a crisis of survival, the installation of public water systems under the name of rationalization may become the final blow to them.

Notes

1 For further discussion refer to Torigoe (1982), which provides one of the most detailed accounts of Aruga's "life analysis" approach. Torigoe (2014) also provides an overview of the approach and its related concepts in English.

2 In this study, "hamlet" (community) refers not only to a geographically delimited area, but also to an organization that its inhabitants formed and maintained over time from necessity to support their livelihoods (Torigoe 1985). This type of geographical community has been often referred to in Japanese rural sociology as a "natural village," as distinguished from an "administrative village" (Suzuki 1940).

3 In Japan there are two major water-supply systems: public and local. Public water systems are maintained by local governments and serve more than 5,000 people. Local water systems are smaller-scale supply systems that serve 5,000 or fewer people. Often local water systems are governed and managed by small communities. Environmental sociologists in Japan have explored the different ideologies underlying the two water systems (e.g. Sakurai 1984; Torigoe 2012).

4 Not included in this table is a household that was using the *Yamanokami* water for drinking and other uses until March 2013. The household head lost his wife and subsequently suffered from a severe illness. As he could no longer participate in the monthly cleanup work, and not wanting to inconvenience other members, he removed himself from the association. A private well was installed for this household using the subsidy system.

References

Aruga, Kizaemon. 1968. *Mura no Seikatsu Soshiki [Life Organizations of Village]*. Tokyo: Miraisha. [In Japanese]

Ikegami, Kouichi. 2007. "Mura ni totte no Shigen towa" [Resources for a Village Community]. In *Mura no Shigen wo Kenkyusuru—Firudo karano Hasso* [Ideas from Fieldwork—A Study of Resources in a Village]. The Japanese Association for Rural Studies (eds), 14–26. Tokyo: Rural Culture Association Japan. [In Japanese]

Ishikawa, Masuo, Yukari Kozako, and Kazunori Suzuki. 2004. Suido Mifukyu Chusankanchi ni okeru Inryousui no Arikata ni tsuite [The State of Drinking Water in a Mesomountainous Region without Public Waterworks—A Study of Water in Kawauchi Village]. *Journal of the Center for Regional Affairs*, Fukushima University 16(1): 5174–5197. [In Japanese]

Kanebishi, Kiyoshi. 2014. *Disaster: Memento Mori—Second Tsunami Wave*. Tokyo: Shinyo-sha. [In Japanese]

Klein, Naomi. 2007. *The Shock Doctrine: The Rise of Disaster Capitalism*. New York: Metropolitan Books.

Sakurai, Atsushi. 1984. "Kawa to Suidou—Mizu to Shakai no Hendo" [River and Waterworks—Change of Water and Society]. In Torigoe Hiroyuki and Kada Yukiko (eds.) *Mizu to Hito no Kankyoshi* [Environmental History of Water and Humanity], 163–204. Tokyo: Ochanomizu Publishing. [In Japanese]

Sakurai, Atsushi. 2014. Chiiki Komyuniti no Seizon Senryaku—Higashi Nihon Daishinsai ni okeru Hisaichi no Taio kara [Survival Strategies for the Local Community in the Great East Japan Earthquake Disaster]. *Journal of Applied Sociology* 56: 1–16. [In Japanese]

Suzuki, Eitaro. 1940. *Nihon Noson Shakai Genri* [Principles of Japanese Rural Societies]. Tokyo: Jicho-sha. [In Japanese]

Torigoe, Hiroyuki. 1982. *Tokara Retto Shakai no Kenkyu* [Study of Tokara Islands Societies]. Tokyo: Ochanomizu Shobo. [In Japanese]

Torigoe, Hiroyuki. 1985. *Ie to Mura no Shakaigaku* [Sociology of Family and Village]. Tokyo: Sekai Shisosha. [In Japanese]

Torigoe, Hiroyuki. 2012. *Mizu to Nihonjin* [Water and People of Japan]. Tokyo: Iwanami Shoten Publishers. [In Japanese]

Torigoe, Hiroyuki. 2014. Life Environmentalism: A Model Developed under Environmental Degradation. *International Journal of Japanese Sociology* 23(1): 21–31.

5 Securing mobility in the nuclear disaster-afflicted region

A case study of Minami-Soma

Itsuki Yoshida

Introduction

Public transportation plays a vital role in securing the mobility of people, especially the socially vulnerable. The Great East Japan Earthquake Disaster seriously damaged not only the livelihoods of individuals and households in the Tohoku region, but also their means of mobility. Following the earthquake and subsequent tsunami disaster, public transportation has played critical yet changing roles, first facilitating the evacuation of residents from the disaster-stricken areas and later securing the mobility of returnees in recovering communities and evacuees shuttling between evacuation and evacuated sites. Needless to say, what has made the impacts of this disaster on public transportation particularly unique and challenging is the nuclear accident at the Fukushima Daiichi Nuclear Power Plant (NPP). The invisible and elusive nature of radioactive contamination resulted in a whole set of novel problems, including the prolonged existence and continuous revision of "restricted areas" (Yamakawa 2016), a large number of so-called "voluntary evacuees" (Horikawa 2016), and the need to deal with the removal and storage of contaminated soil (Fujimoto 2016). These problems all have indirect bearings on the local public-transportation infrastructure and workforce. Yet they receive hardly any discussion in the existing literature on the effects of large-scale disasters on local public transportation (e.g. Chang and Nojima 2001; Chang 2003; Litman 2006) due to the rarity of massive nuclear disasters.

This chapter examines the transformation of public transportation in the disaster-afflicted areas. I focus on Minami-Soma City, a municipality that has been significantly affected by both the earthquake/tsunami and the nuclear accident due to its proximity to the nuclear power plant. Unlike other towns even closer to the plant, such as Futaba and Okuma, Minami-Soma has had its evacuation orders gradually lifted. This chapter demonstrates that while some of the post-disaster responses and challenges regarding local public transportation stem from the peculiar nature of a nuclear accident and radioactive contamination, others must be seen as manifestations of pre-disaster socio-institutional arrangements.

Minami-Soma City: an outline

Minami-Soma City is located on the coast of Fukushima Prefecture, about 10–40 km north of the Fukushima Daiichi NPP. It is made up of three main wards: Odaka, Kashima, and Haramachi (Figure 5.1). The population of the city at the time of the disaster was 71,651, making it the largest municipality included within the "restricted

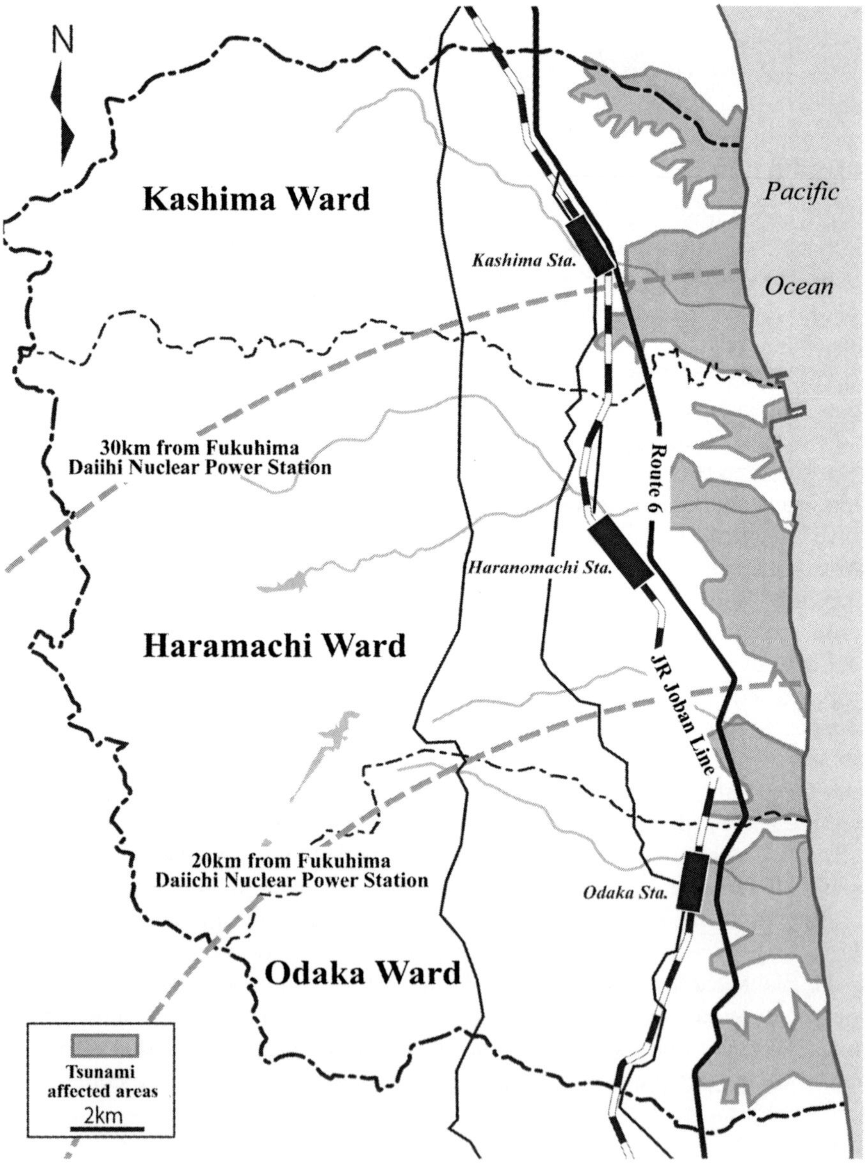

Figure 5.1 Map of Minami-Soma.

Source: Yoshida *et al.* (2012).

areas" designated after the accident. During the earthquake, the city experienced massive shaking, and residential districts, agricultural land, and the fishing port in the eastern coastal areas of the city were devastated by the tsunami that followed. As of March 11, 2014, total deaths resulting from the disaster, including so-called earthquake-related deaths, numbered 1,083, the highest for any municipality in Fukushima Prefecture. As of March 15, 2012, at least 3,655 households had reported damage to their residences, varying from partial damage to total destruction.

In addition to the damage from the earthquake and tsunami, Minami-Soma City was also severely affected by the nuclear accident. Figure 5.2 presents a timeline of the events of the nuclear accident and the transformations of the "restricted areas" and "nuclear disaster-afflicted areas" from the time of the accident until the present day. On March 12, 2011, residents of the Odaka ward—an area located mostly within a 20 km radius of the Fukushima Daiichi plant—were evacuated. On March 15, residents within a 20–30 km radius of the plant, including the central business and administrative districts of Minami-Soma City, were ordered to stay indoors. While this indoor-sheltering order was not meant to spur evacuation within or outside the city, fears of radiation contamination impeded the flow of people and goods into the area, inducing many residents to evacuee "voluntarily" (Yamakawa 2016; Horikawa 2016). This stoppage of inward flows continued

Extent of Minami-Soma City

Date and time	Distance from Fukushima Daiichi Nuclear Power Plant: – 3km	3–10km	10–20km	20–30km
21:23, March 11	Evacuation order	Indoor sheltering order	No designation	
5:44, March 12	Evacuation order			
15:36, March 12	Reactor 1 exploded			
18:25, March 12	Evacuation order			No designation
11:01, March 14	Reactor 3 exploded			
6:10, March 15	Reactor 2 exploded			
11:00, March 15	Evacuation order			Indoor sheltering order
April 22–Sept 29, 2011	Restricted Area			Deliberate Evacuation Area; >20mSv/y Evacuation-prepared Area
Sept 30, 2011–April 15, 2012	Restricted Area			Deliberate Evacuation Area; >20mSv/y No designation
April 16, 2012–	Restricted Area		Difficult to Return Area; >50mSv/y Limited Residence Area; >20mSv/y Evacuation Lift Preparation Area; <20mSv/y No designation	

Figure 5.2 Timeline of evacuation orders and evacuation zones.

Source: Yoshida (2012).

until April 22, when the indoor-sheltering order was lifted and the area was partitioned into various afflicted-area designations.

Table 5.1 presents annual data pertaining to the post-disaster residential locations of the 71,651 residents who lived in Minami-Soma City at the time of the disaster. More specifically, these data indicate the number of residents who have returned, remained in evacuation, or moved away from the city. As evidenced by the data, even as of August 2015—four years and five months after the disaster—around 16 per cent of the population (11,188 individuals) remained in evacuation outside the city, while only approximately half of the population (34,888 individuals) had returned to their former homes. Moreover, while the number of residents who have opted to move away from the city has increased, evacuees from areas closer to the nuclear power plant (such as those from Okuma and Namie Towns) have moved into Minami-Soma. However, since part of Minami-Soma remains designated as a nuclear disaster-afflicted area (and evacuation orders remain in effect), overnight stays within those areas are still prohibited. Accordingly, approximately 8,000 citizens are living in emergency temporary housing and government-subsidized rental units in the city. Some evacuation orders will certainly be lifted in the future, but it is impossible to predict how many citizens will return to their former homes and neighborhoods. The magnitude of the earthquake disaster, the prolonged and continuous reconfigurations of the nuclear disaster-afflicted areas, and the uncertainties arising from radioactive contamination all continue to pose great difficulties for local transportation operation and planning. To explore these challenges further we need to consider the specific effects of the disaster on local public transportation, and how public transportation has been used to respond to the changing phases and circumstances of the disaster.

Table 5.1 Yearly population by residence type in the city of Minami-Soma

As of:	*Aug. 11*	*Aug. 12*	*Aug. 13*	*Aug. 14*	*Aug. 15*
1 Minami-Soma City	38,557	44,959	46,607	47,256	47,334
Own houses, original location	31,797	34,588	35,239	35,270	34,888
Shelters	196	0	0	0	0
Temporary houses	2,785	3,227	5,535	5,166	4,243
Rented houses etc.*	3,779	5,032	5,833	3,786	3,725
Own houses, new location	–	–	–	3,034	4,478
2 Other municipalities	27,181	19,383	15,431	13,169	11,188
Shelters	4,123	2	0	0	0
Temporary houses	0	0	0	0	0
Rented houses etc.	23,058	19,381	15,431	13,169	11,188
3 Moved out	3,269	5,272	6,871	7,823	8,924
4 Dead	1,008	1,756	2,549	3,292	4,086

Source: Reconstruction Planning Department of Minami-Soma.

Note: "Rental houses etc." included own houses in new locations until August, 2013.

Changes in mobility after the disaster

Before the earthquake, transportation services in Minami-Soma City included the Japan Railways (JR) Joban line (five stations); 24 bus lines operated by three separate companies; on-demand transportation[1] known as the "Odaka e-machi taxi" in the Odaka ward; and school buses for some elementary schools and hospital shuttle buses in the Kashima ward. Some of these local public-transportation services were paralyzed following the disaster, but others provided essential means of emergency evacuation. Subsequently, new transportation services emerged to meet the changing demand brought by shifts in the designations of nuclear disaster-afflicted areas and in other parameters of the disaster. Accordingly, I document below the shift in mobility needs and adaptations following the accident and various evacuation designations over three successive periods: 1) the "emergency period" from the disaster on March 11, 2011 until the designation of nuclear disaster-afflicted areas on April 22, 2011; 2) the "transition period" from then until the emergency evacuation preparation area designations were lifted on April 15, 2012; and 3) the "recovery period" that has followed.

The emergency period (March 11–April 22, 2011)

During the first month of the disaster, securing the mobility of evacuees was the most important task. The public-transportation services within Minami-Soma City were suspended in the immediate aftermath of the earthquake disaster. However, following the nuclear accident, from March 15–24, a temporary evacuation was implemented. Using buses provided by the Minami-Soma municipal government, 5,000 residents evacuated in groups headed for areas within and beyond the prefecture. However, because residents received conflicting signals about the status of the plant and the accident, many out-of-town buses refused to enter the city limits. Accordingly, transit points were established near the city's border with neighboring Kawamata Town and Nihonmatsu City. Bus operators within the city as well as buses owned by the municipal government shuttled residents from the city to these transit points, from which outside buses shuttled residents to evacuation destinations. It is important to note that these were not predetermined evacuation plans, but rather *ad hoc* strategies adopted by each individual municipality after being ordered to evacuate.

In addition, while most areas of the Haramachi ward of central Minami-Soma were placed under indoor-sheltering designations, many residents opted to evacuate on their own, and the offices and vehicles of public-transportation operators were also temporarily relocated outside the city. Accordingly, vehicles for evacuation were not easily procured and since most residents evacuated using their private vehicles, serious traffic backups were experienced on the area's main roads. As a result of this evacuation, it is estimated that Minami-Soma's approximate population of 70,000 residents before the earthquake disaster had decreased to 10,000 by the end of March 2011 (interview with the staff of the Reconstruction Planning Office of Minami-Soma City, April 9, 2012).

The transition period (April 22, 2011–April 15, 2012)

Once the initial crisis situation had abated, the focus of transportation concern shifted to linking Minami-Soma with major cities and to providing the most important services for the returnees, primarily by bus. With the designation of "nuclear disaster-afflicted areas" on April 22, 2011, the indoor-sheltering orders were lifted for areas within a 20–30 km radius of the nuclear power plant. Subsequently, these areas were re-designated as either "planned evacuation" or "emergency evacuation preparation" areas based on air radiation-dose rates. Three elementary schools and one junior-high school located in the Kashima ward outside the 30 km radius of the nuclear power plant were re-opened, and a temporary combined elementary and junior-high school constructed in the Kashima ward also began operation. Amid this evolving situation, residents of the Haramachi ward, an area that includes the central business and administrative districts of the city, began to return to their homes. Accordingly, in addition to restarting the operation of a local bus service connecting residents to the main hospital in the area, Minami-Soma City commissioned a local operator to provide school-bus services for the elementary and junior-high school in Kashima ward. Because recovery of the Japan Railways line that connects this area of the city with Sendai to the north and Tokyo to the south was still not on the horizon, an intercity bus was established to link Minami-Soma with Sendai and Fukushima City. The establishment of such intercity bus services was not limited to Minami-Soma or the nuclear disaster-afflicted areas, but also frequently occurred in the earthquake and tsunami-devastated areas.

An important characteristic of these intercity bus services was that they were established by private operators. They avoided high-dose areas and impassable roads, and were also a key means of transportation into the area for relief workers and for evacuees to make short visits during the transition period. However, at this stage, bus lines serving most medical and commercial facilities in the area remained largely out of operation. This is partly because the principle that bus operators will provide services based on their own assessment of costs and benefits is deep-seated, and Minami-Soma was no exception to this rule. So long as many residents remain in evacuation outside the city and medical and retail facilities remain closed, buses operating only within the city cannot be expected to be profitable. Additionally, since many bus-operating personnel have also evacuated, bus operators prioritize commissions to establish school buses and new intercity lines where demand is reliable.

On September 30, 2011, following the lifting of the "emergency evacuation preparation area" designation that had covered most of Haramachi ward, schools in this ward were successively re-opened. Additionally, emergency temporary housing was established, primarily within Kashima ward, and efforts were made to connect this temporary housing to medical and retail facilities. Shuttle buses to the Kashima Health and Welfare Center that had resumed during the "emergency period" opened new additional lines. Additionally, using the grant system established to support the recovery of the Fukushima region, Minami-Soma City contracted with Haramachi Travel, a local transportation company, to operate two temporary bus lines in the city, with each new line operating three days a week. In this way, as the number of evacuees returning to Minami-Soma began to increase with the continuous redefinition of evacuation zones, public-transportation services were made available to provide for their needs. However, these services were mainly school buses and transportation from emergency housing; the recovery of pre-disaster local bus services has not progressed. A collaborative questionnaire survey[2] conducted by the Reconstruction Planning Department of Minami-Soma and the present author asked residents of Minami-Soma to compare their mobility before and after the disaster. Residents residing in evacuation sites felt "more reluctant to make trips" compared to individuals living elsewhere (most presumably live in Minami-Soma). However, there were more individuals living in Minami-Soma who were actually making fewer trips, compared to those living in evacuation sites (Table 5.2). Thus, securing the mobility of the returnee residents in their daily lives will be an important task in the near future.

Additionally, in order to respond to the strong demand for some means of transportation for commuting to high schools, on December 21 a section of the Japan Railways Joban line that had been out of operation was re-opened between Soma and Hara-no-Machi stations. With regard to inter-regional transportation, the number of buses to Fukushima City and Sendai were increased, and an alteration in bus routes to bypass Soma City shortened the length of bus trips to Fukushima.[3]

The recovery period (April 16, 2012–present)

One year after the onset of the nuclear disaster, securing links between evacuees and their homes in afflicted areas became an important mobility objective. On April 16, 2012, restricted areas in Minami-Soma were re-designated. Most of the Odaka ward became evacuation-lift preparation, or limited-residence, areas (Figure 5.2). In both of these zones, evacuation orders remained in place and overnight stays remained prohibited. However, residents were now able to return without screening or radiation dose-management precautions. For that reason, demand increased

Table 5.2 Changes in trip-related behaviors after the nuclear accident

	Forms of housing		*p-value*
	In evacuation housing	*Other housing*	
Making fewer trips	33.3%	41.4%	0.01 *
More reluctant to make trips	33.3%	27.4%	0.04 *
Number of respondents	297	1,186	

*$p < 0.05$

Source: Collaborative questionnaire survey conducted by the Reconstruction Planning Department of Minami-Soma and the author.

for transportation to these areas in order to conduct preparations for returning, such as cleaning and repairing houses, as well as for managing family gravesites and altars. However, after these areas were re-designated, transportation services to the Odaka ward still remained out of operation. It subsequently became known that many of the individuals attempting to return to the re-designated areas were older people without their own cars, who were transported by friends and relatives (interview with the staff of the Reconstruction Planning Office of Minami-Soma City, August 10, 2012). In response to this situation, Minami-Soma City contracted with the Fukushima Future Center for Regional Revitalization (FURE) at Fukushima University to conduct a "temporary return transportation project." Fukushima University subsequently hired two companies (Sanwa Shoukai and Fuji Taxi) that had been operating on-demand transportation services within the Odaka ward to extend their services to and from the evacuated areas. On October 30, 2012, operation of the "Jumbo Taxi," a full-sized shuttle van, commenced (Figure 5.3).

The Jumbo Taxi is a three-day-per-week, on-demand transportation service. Passengers make reservations in advance and the Jumbo Taxi picks them up at temporary-housing bus stops where stops have been scheduled, transporting them to their houses in the former restricted areas. From the beginning of operations up to January 2015, this service has been extremely popular, providing 3,290 rides to 132 different individuals. Moreover, 17 individuals have used the Jumbo Taxi for more than 50 different return trips to their residences. While demand for public transportation in normal times usually arises for "activity-based trips" such as commuting to work or school, shopping, or hospital visits, in the restricted areas serviced by the Jumbo Taxi there is no such demand because such facilities

Figure 5.3 Operation of the "Jumbo Taxi".

Photo credit: The City of Minami-Soma.

remain out of operation. However, the fact that there are evacuees who regularly use the Jumbo Taxi suggests that it not only meets their need for temporary return visits but also has a social value for the maintenance of pre-disaster community ties, which have been under stress due to the dispersed evacuee-housing characteristic of this disaster (Chapter 1).

Adaptive responses of bus operators after the disaster

In contrast to the slow recovery times for railways following a large-scale disaster, passenger-bus lines can be restored comparatively quickly, and since routes and frequency can be easily adjusted, from the perspective of the mobility of the afflicted areas they are a vital mode of transportation. In this section I compare the response of bus operators in Minami-Soma with that in Ofunato City, Iwate Prefecture, which was damaged severely by the tsunami, but not directly by the nuclear accident. Through this comparison I distinguish local and national factors that influenced the bus operators' responses.

Table 5.3 presents data pertaining to the damage to and recovery of bus operations in Minami-Soma City and Ofunato City. There are a number of similarities between Ofunato and Minami-Soma. One is that local bus routes within these cities were only restored well after express-bus services to Morioka and Sendai were restored. In Ofunato, for example, the express-bus service between Ofunato and Morioka was reestablished on March 19. Moreover, in Minami-Soma, Haramachi Travel Ltd.,[4] a bus operator that had not operated intercity buses before the accident, established a new route to Sendai City and was also able to furnish transportation earlier and more effectively than the substitute services offered by the Japan Railways Joban line. Second, in both cities the government commissioned transportation services that had not been in operation prior to the disaster. In Ofunato this included transportation to bathing facilities for evacuees provided by U.S. military rescue teams among others, and in Minami-Soma it included school buses to re-opened elementary and junior-high schools. Third, in both cities the local bus lines that support commutes to school and work, hospital visits, and shopping began to be restored, after intercity buses. Moreover, transportation services within the local area were significantly altered in a piecemeal fashion to respond to the urgencies of the recovery phase, as evidenced in the use of buses to help with relief efforts and to transport residents to bathing facilities.

The first point of similarity, the early restart of the express buses and the opening of new lines, is particularly noteworthy. My analysis of the post-disaster Tohoku-wide transportation recovery[5] indicates that this phenomenon was partially spurred by response measures adopted by the Ministry of Land, Infrastructure, Transport and Tourism (MLIT). To provide means of transportation for the disaster-afflicted areas, and especially a substitute for the out-of-operation Tohoku Shinkansen, charter-bus operators, who are usually not allowed to operate on regular routes, were given special permission to operate new temporary inter-regional services. Additionally, the subsidy requirements of the Securing and Maintaining Local Public Transportation Project—a government subsidy system

Table 5.3 Damage from the disaster and restoration of service by local public bus operators in Ofunato and Minami-Soma

City	*Bus operator*	*Damage from the disaster through March 2011*	*Status of restoration by April 2011*
Ofunato (Iwate Pref.)	Iwate Bus Co., Ltd.	Mar. 11: The terminal is swept away by tsunami. Mar. 13: The terminal is temporarily restored at a company-owned plot on a hill. Mar. 19: Express intercity bus service to Morioka is restored.	Apr. 4: Scheduled bus service in the city is restored. Apr. 22: Intercity service to Rikuzentakata is restored. Apr. 28: Express intercity bus service to Sendai is restored.
Minami-Soma (Fukushima Pref.)	Fukushima Transportation, Inc. Shin Joban Kotsu Co., Ltd. Haramachi Travel, Ltd.	After the nuclear accident, each bus operator relocates its bus terminal to outside the city.	Apr. 15: Haramachi/Sendai intercity bus service is inaugurated. Apr. 22: Scheduled bus service within the city is restored. Apr. 27: Soma/Haramachi intercity service is restored.

Source: Yoshida (2015).

for fixed-route buses—have been relaxed through the 2015 budget for disaster-afflicted municipalities. Additionally, even charter buses operating intercity bus services are now eligible for these government subsidies.[6] This system proved effective for securing inter-regional transport after the disaster, and is a policy that should be considered again in the case of another large-scale disaster.

We now turn to some noticeable differences between the two cities. As noted earlier, in Minami-Soma, even now, evacuation designations remain in place and more than a few bus lines have not been restored. This is in stark contrast to Ofunato, where all lines in the city have been restored even though the central business district experienced massive damage and the office of Iwate Transportation, a transportation company operating within the city, was washed away. While this can be seen as connected to the unique characteristics of a nuclear disaster, it is also related to different levels of local-government involvement and to prior experiences in local transportation.

First, neither city put forth much effort to provide public transportation prior to the earthquake. In Ofunato, however, from the initial phase of emergency temporary-housing provision, in addition to proactively negotiating with bus operators on how to secure mobility, the city sought increased connections with emergency temporary housing and the restoration of inner-city bus services in tandem. On the other hand, in Minami-Soma, since school buses and routes to emergency temporary housing were provided largely on a case-by-case basis as needs arose, it was not possible to efficiently deploy resources such as vehicles and drivers, and this became one important factor in the slow recovery of bus lines. Thus, we can see that in order to secure mobility after a large-scale disaster, when the human resources of bus operators and government are scarce, it is important to deal with local public transport as an adaptive network. If such planning is only initiated after a disaster has occurred, responses will certainly be delayed. In addition to strengthening collaboration between government and bus operators even during normal times, it is necessary to decide on policies for securing public transportation in the case of a disaster, which will also include transportation services provided by non-profit organizations (as seen in the case of the Jumbo Taxi).

Second, prior experiences by the transportation companies were also important. In Ofunato City, of the 31 buses owned by Iwate Transportation, nine were washed away by the tsunami and heavily damaged. The office staff and drivers at the offices evacuated their buses to a property on higher ground owned by the company, and thus significantly limited damage to the bus fleet. Although a tsunami evacuation and response manual had not been prepared, and was not available for reference at these offices, a tsunami warning resulting from an earthquake in Chile during the previous year had spurred an evacuation of areas of Iwate Prefecture along the coast, and the staff and drivers of Iwate Transportation therefore had some experience that proved useful and effective during this disaster (Interview at the Ofunato office of Iwateken Kotsu Co., July 28, 2011). Indeed, only two days after the initial disaster, following a request from the Ofunato municipal government, Iwate Transportation re-established their offices at a bus depot in Takkon Town. In the initial aftermath of the disaster, Iwate Transportation offered transportation assistance to rescue efforts conducted by United States military forces stationed in Japan and also operated bus services between evacuation centers and bathing facilities opened by the Japan Self-Defense Forces in late March.

In contrast, bus operators in Minami-Soma, also without evacuation manuals, lacked the kind of experience that the Ofunato bus operators had, and this may have played a role in shaping their response at least during the emergency period immediately after the disaster. Of course, virtually no transportation operators in Japan had any experience of, nor did they anticipate, a nuclear disaster of this scale. It is therefore critical to examine, especially during the recovery period, what unique conditions exist in nuclear disaster-afflicted areas, as I discuss next.

Securing mobility in the nuclear disaster-afflicted areas

Several factors must be adequately taken into consideration in securing mobility in the nuclear disaster-afflicted areas. First is the dynamic geography of evacuees and returnees. As noted above (see Table 5.1), even as of August 2015, approximately 11,000 evacuees from Minami-Soma City remained in evacuation both within and beyond the prefecture. Additionally, great contrasts in evacuation conditions are found within the city. For example, while the entire Odaka ward area remains under evacuation and the population is thus zero, in the neighboring Kashima ward, where temporary housing has been constructed on a large scale, the current population of over 13,000 now exceeds the pre-disaster population of 11,603. These shifts in the spatial distribution of the population matter greatly when formulating measures to provide mobility to the disaster-affected areas. In the case of Minami-Soma, it is likely that many former residents in evacuation will stay away from, or only temporarily visit, their original homes for a long time. In addition, because only part of the city is designated as a forced evacuation zone, the municipal office functions have returned. But because this situation makes it difficult for the many evacuees of Minami-Soma residing outside the city to become involved in community-planning efforts, the regenerative strength that must serve as the basis of local recovery and reconstruction risks being squandered. Accordingly, even more than in normal times, it is essential to ensure mobility across the city and region.

The second factor pertains to the issue of interim facilities to manage and store the contaminated soil removed during decontamination. On September 1, 2014, the governor of Fukushima Prefecture announced plans to permit the construction of these facilities. Looking ahead, the prefecture has asserted that decisions to permit construction and to allow materials arising from decontamination to be transported to these facilities are separate ones that must be considered in succession.[7] Accordingly, while ensuring the safety of the interim storage facilities it will also be imperative to guarantee the safe and smooth transport of the contaminated soil distributed throughout the prefecture. Since trucks carrying contaminated materials will have to pass close by residences and other road users, there are concerns about the traffic congestion that this concentration of vehicles will generate. Accordingly, as much as is possible, efforts should be made to ensure that the routes of these trucks are spatially and temporally separated from regular traffic through the use of special routes, and also by using railway transportation when feasible.

The third factor is the potential shortage of bus drivers, which is an outgrowth of the interim storage problems. Assuming that 22,000,000 m^3 of contaminated soil will be hauled over the course of three years (based on 250 annual working days), the Ministry of the Environment estimates that 1,500 to 2,000 ten-ton trucks (including dump trucks) will be required (Fukushima Headquarters for Interim Storage Facilities 2013). As of March 2014, 2,329 ten-ton trucks were registered in Fukushima Prefecture, meaning that more than half of all the ten-ton trucks in the prefecture, and their specially licensed drivers, will need to be allotted to this task. There are concerns that competition will arise between decontamination operations and transportation providers over the limited number of specially licensed drivers in the area. Many of them will likely opt to drive the freight trucks in high demand for reconstruction and decontamination purposes in order to gain higher wages, accelerating the current shortage of qualified drivers for the transportation network. Furthermore, as indicated in Table 5.4, the number of individuals holding large-sized motor vehicle second-class licenses—special licenses for operating buses that are a key human-resource component for a public-transportation network—in Iwate, Miyagi and Fukushima Prefectures is decreasing and, as evidenced by the fact that the average age of these holders is over 60, the personnel holding these special licenses are also aging. Given that average earnings of employees of private bus operators were 920,000 yen lower than average earnings across all industries in 2008, many holders of special licenses opt to drive trucks, which are now in high demand for reconstruction and decontamination work. The shortage of bus drivers will remain a serious issue in coming years. Thus, while at first glance they may appear to be separate issues, the need to secure drivers to transport contaminated materials is actually closely linked to the sustainability of transportation service provision in the area.

Last, there are also many remaining future tasks related to the still only partially recovered and reconstructed sites that a transportation network is supposed to connect. In particular, the slow recovery of retail businesses can be attributed to a decline in the number of shoppers, shortages of full- and part-time workers, and the fact that suppliers are out of operation. This situation, in turn, has further decreased accessibility for residents, and this is another issue that needs to be tackled in the future.

Conclusions: future tasks for mobility in the disaster-afflicted areas

This chapter used a case study of Minami-Soma City to outline transformations in the conditions of evacuation life and mobility over time that have resulted from the earthquake disaster and the nuclear accident. One of the defining features of

Table 5.4 Changes in the number of large-size motor vehicle, second-class license holders and their average ages

	2008		*2010*		*2012*	
	Licensee	*Average age*	*Licensee*	*Average age*	*Licensee*	*Average age*
Iwate	12,519	60.3	12,388	60.4	12,071	60.8
Miyagi	21,253	58.9	20,961	59.6	20,744	60.2
Fukushima	18,421	59.9	18,191	60.2	17,836	60.8

Source: Compiled by the author (Yoshida 2014) based on *Drivers License Statistics* by the National Police Agency.

the evacuation of Minami-Soma City is that evacuees are living at a great distance from their former homes, often beyond the city and beyond the prefecture. The dispersed nature of the evacuation makes the provision of a spatially extensive transportation network an absolute necessity. With full recovery of the region's railway network not yet on the horizon, the actions taken by former and new bus operators to provide intercity bus routes have proved highly effective in securing the extensive transportation network necessitated by dispersed evacuation.

The adaptive transportation responses pursued in Minami-Soma City—and indeed in all of the areas affected by the nuclear accident and the earthquake and tsunami—have focused on securing mobility to emergency temporary housing. However, several years after the disaster, it is not entirely clear whether local public-transportation networks can be secured and maintained, even with the continuation of the generous subsidies and policies that have supported these networks until now. This is because evacuees will at some point in the future eventually return to permanent housing and, amid the highly problematic population decreases and aging of the disaster-afflicted municipalities along the coast,[8] just as in normal times, each municipality (or each residential neighborhood) will be required to develop policy guidelines for local public transport and strategically plan to secure, maintain, and improve local public-transportation networks.

As a result of the deregulation of bus operations in February 2002, responsibility for managing non-profit local public transportation was left to regional governments, and municipal governments in particular. Both the Local Public Transportation Conference System established after the revisions to the Road Transportation Law of October 2006 and the Act on Revitalization and Rehabilitation of Local Public Transportation Systems implemented in October 2007 were attempts to break away from a system that left local public-transportation decisions to private operators and move towards securing, maintaining, and improving local public-transportation networks through the formation of consensus between government, transportation operators, and local residents. In addition, the Basic Act on Transport Policy enacted on December 4, 2013 emphasizes the agency of local governments in formulating and implementing transportation policies according to locally specific environmental characteristics and socioeconomic needs.

However, in the earthquake and nuclear disaster-afflicted municipalities along the coast, the Local Public Transport Meeting System is either not yet established or not functioning; plans for local public transport are not yet established; and there are many cases where frameworks for strategic planning to secure, maintain, and improve the local public-transport network are not in place. Accordingly, while reconstruction plans for each community contain maps and diagrams for local infrastructure and facilities, these plans contain little consideration of how residents will actually move through these newly planned landscapes. Without further consideration of public transport, even when public-reconstruction housing and higher ground-relocation housing is constructed and evacuees return permanently, it is highly likely that, without their own automobiles or without someone to offer them a ride, residents' essential tasks of daily life will be significantly constrained. Accordingly, it is necessary to develop policy guidelines for local public transport and to conceptualize a transportation network from the initial phase of reconstruction-oriented community-based planning, or what the Japanese call *machi-zukuri* (town-making). In the disaster-afflicted municipalities along the coast, buses and other public-transportation services were at a low level—with few routes and infrequent buses—even before the earthquake, and residents were highly dependent on personal automobiles. If an infrastructural environment that enables citizens to freely move about is not created, then chances for interaction between fellow citizens will be lost. That interaction is necessary for community-building and for re-activating the local economy and revitalizing the central commercial areas of the city.

In short, excluding the emergency responses of the period that immediately followed the disaster, many of the issues and tasks related to mobility that have come to the fore as a result of this disaster were not novel issues. They were challenges that were already present, but accentuated and made more prominent by the disaster. Thus, it would seem no exaggeration to say that the quality of transportation policies in normal times greatly influences the course of responses to a large-scale disaster.

Acknowledgment

This work was supported by JSPS KAKENHI Grant Number 25220403.

Notes

1 On-demand transportation is a form of transportation in which users make reservations in advance and routes and schedules are then tailored to meet their needs.

2 This survey was conducted between July 23 and August 12, 2015 with a randomly selected sample of 3,072 households. Responses were received from 1,031 households (1,594 individuals), a response rate of 33.6 per cent.

3 The shortest route between Minami-Soma City and Fukushima City passes through the village of Iitate, but since this village was designated a "planned evacuation area," bus operators established a route that passed outside Soma City.

4 Haramachi Travel Ltd. is now known as Tohoku Access Co.

5 In regard to transportation between the Tohoku and Tokyo regions, until all lines of the Tohoku Shinkansen were restored on April 29, bus operators were able to secure transportation capacity by operating on the same time schedule but increasing the number of buses in operation. As a result, between March 26 and April 1, the third week after the accident, a daily average of 7,355 people were transported on 31 routes (Ministry of Land, Infrastructure and Transport and Tourism (MLIT) presentation materials, May 16, 2011). Since bus operators' pre-accident transport capacity consisted of 30 routes serving 1,980 individuals per day (one bus per time slot), even if buses were operating at full capacity before the earthquake, post-disaster ridership was four times the previous capacity.

6 Within the Securing and Maintaining Local Public Transportation Project, requirements for eligibility for subsidies to support new lines include—among criteria of passenger capacity, for example—that the service must offer transport for more than 15 and less than 150 passengers. However, these criteria are not well adapted to the specifications for inter-regional transportation made annually by regional transportation bureaus.

7 Fukushima Prefecture has also stipulated to the national government five conditions that must be fulfilled before any final approval is given to transport materials to these facilities, including continuing maintenance of the transportation routes and clarification of policies for the surrounding areas, and also ensuring the safety of the facilities and transport (Governor of Fukushima 2014). In response, on March 13, 2015, transporting testing operations were initiated to move materials to depots constructed in Okuma Town and Futaba Town. Contaminated soil removed during the ongoing decontamination process is scheduled to eventually be transported to interim storage facilities, but is currently being stored at temporary facilities and on-site at decontamination sites. It has been pointed out that there are many critical issues at stake with these plans, due in particular to the sheer volume of the 22,000,000 m^3 of soil that is to be removed and the vast expanses across which this soil must be collected and transported.

8 For example, in the case of Minami-Soma City, the per centage of the population over 65 years of age before the earthquake disaster in February 2011 was 25.9 per cent; two years later, in January 2013, this per centage had increased to 28.3 per cent. Even with many people moving away from the area as a result of the nuclear disaster, the elderly population has remained largely unchanged (as evidenced by an increase of 32 individuals in the above comparison), further increasing the per centage of the population that is over 65.

References

Chang, Stephanie E. 2003. "Transportation Planning for Disasters: An Accessibility Approach." *Environment and Planning A* 35(6): 1051–1072.

Chang, Stephanie E., and N. Nojima. 2001. "Measuring Post-Disaster Transportation System Performance: The 1995 Kobe Earthquake in Comparative Perspective." *Transportation Research Part A: Policy and Practice* 35(6): 475–494.

Fujimoto, Noritsugu. 2016. "Decontamination-Intensive Reconstruction Policy in Fukushima Under Governmental Budget Constraint." In *Unravelling the Fukushima Disaster*, edited by Mitsuo Yamakawa and Daisaku Yamamoto, pp. 106–119. London: Routledge.

Fukushima Headquarters for Interim Storage Facilities. 2013. *Jyokyo Dojou tou no Chuukan Chozou Shisetsu no An ni tsuite* [Interim Storage Facilities Plan for Contaminated Soil], Fukushima: Reconstruction Agency. [In Japanese]

Governor of Fukushima. 2014. *Chuukan Chozou Shisetsu ni kakaru Moushiire ni tsuite* [Matters Relating to Interim Storage Facilities], Fukushima: Fukushima Prefecture. [In Japanese]

Horikawa, Naoko. 2016. "Displacement and Hope After Adversity: Narratives of Evacuees Following the Fukushima Nuclear Accident." In *Unravelling the Fukushima Disaster*, edited by Mitsuo Yamakawa and Daisaku Yamamoto, pp. 66–78. London: Routledge.

Litman, Todd. 2006. "Lessons From Katrina and Rita: What Major Disasters Can Teach Transportation Planners." *Journal of Transportation Engineering* 132(1): 11–18.

Yamakawa, Mitsuo. 2016. "Living in Suspension: Conditions and Prospects of Evacuees from the Eight Municipalities of the Futaba District." In *Unravelling the Fukushima Disaster*, edited by Mitsuo Yamakawa and Daisaku Yamamoto, pp. 51–65. London: Routledge.

Yoshida, Itsuki. 2012. "The Study of Providing Mobility and Improving Individual Accessibility in a Disaster Area of the Great East Japan Earthquake." *Transport Science* 34(1): 11–18. [In Japanese]

Yoshida, Itsuki. 2014. "Current Status and Problems on Local Transport Services in the Disaster Area of the Great East Japan Earthquake." *The Toshi Mondai* [The Municipal Problems] 105: 52–62. [In Japanese]

Yoshida, Itsuki. 2015. "Changing Situations and the Issues of Mobility in Disaster Areas of Japan." *Proceedings from the International Conference on Mobility and Transport for Elderly and Disabled Persons* 14: B93–B108.

Yoshida, Itsuki, Katsuyuki Matsuura, Kenji Kawasaki, and Jun Hasegawa. 2012. "A Study on Improving Local Mobility after the Great East Japan Earthquake." *Proceedings of Infrastructure Planning* 45: CD-ROM. [In Japanese]

6 Toward effective radioactivity countermeasures for agricultural products

Hideki Ishii

Introduction

As a result of the Fukushima Daiichi nuclear accident, Fukushima Prefecture and localities throughout eastern Japan experienced widespread radiation contamination. Even today, five years after the initial accident, evacuation orders remain in place for many communities. In areas where residents remain, the identification of agricultural products over the set limits for radiation in food continues to lead to restrictions on shipping and distribution. While most agricultural products from Fukushima no longer show levels of radioactivity above the safety limits set by the government, consumers remain suspicious of products from the area. Such stigmatization (*fuhyo higai* or "reputational damage") has led to a seriously challenging state of affairs for retailing Fukushima products. Against this backdrop, farmers are confronted with great adversity. They are forced to question whether it is possible to produce safe and reliable agricultural products, and whether they are safe from external radiation exposure while cultivating their fields. Burdened with such heavy concerns, it should come as no surprise that many individuals have left farming behind (Kimura and Katano 2014). However, over the course of the past five years, important advances have also been made. For example, our understanding of the circulation of radioactive cesium in the environment, as well as techniques for inhibiting the transfer of radioactive cesium, has been greatly advanced (Kawatsu, Ohse, and Kitayama 2016). While a number of research tasks remain in need of further elucidation, it deserves to be noted that, in a relatively short span of time, a great amount of knowledge has accumulated and important new findings are continually being made. Based on our research conducted at Fukushima University, and focusing in particular on the case of rice cultivation, this chapter outlines a system for producing safe agricultural products even under conditions of radioactive contamination.

Differences between production policies and postharvest food inspection

The limits for radiation in food were temporarily increased to 500 becquerel per kilogram (Bq/kg) in 2011, but then lowered to 100 Bq/kg in 2012 in accordance with the Food Sanitation Law.[1] In areas of Fukushima Prefecture where radioactive contamination has been prominent, even in 2013 and 2014 there was a possibility that foods such as *yuzu* (*Citrus junos*) and chestnut would exceed the 100 Bq/kg limit. However, almost all other products were from that time found to be under the limit. Indeed, radiation is not detected in the vast majority of agricultural products—a fact indicative of the degree to which the safety of agricultural products from Fukushima has been established on a holistic basis.

There are two stages at which food is ensured to remain under the limits for radiation: production and post-production. At the former stage—in policies concerned with countering radiation during agricultural production—there are essentially two main branches of radiation countermeasures: zoning based on maps of radioactive contamination to select appropriate crops for particular fields, and measures to control radiation absorption during the process of cultivation (i.e. control of the productive environment including soil, water, feed, etc.). In the case of the latter stage—post-production inspection of radiation in food—there are again essentially two main branches of measures to be taken: monitoring inspection by the national government, prefectures, and the distribution industry, and voluntary testing by consumers at the point of consumption. Voluntary testing by consumers refers to measuring radiation in food products that do not enter the market, such as food cultivated in home gardens, that received as gifts, or mushrooms and wild edibles harvested from forests and fields. Such voluntary testing activities by consumers are significant in terms of supplementing and validating national and prefectural monitoring inspection, and are essential in enabling consumers to procure food products without concern.

Food testing and policies at the stage of production are inseparable. Any decision to emphasize one over the other must be based on consideration of the methods of production for each specific food product. Figure 6.1 provides an illustration of model distributions for the concentration of radioactive materials in food. In this graph the horizontal axis represents the concentration of radioactive materials in food, while the vertical axis represents the total amount of food. The bell-shaped curves model the distribution of a single food with varying levels of radioactive contamination. The statistical distribution of food with varying levels of radioactive contamination prior to the implementation of countermeasure policies is represented by the solid curve. If the government-set limit for radiation in food is 100 Bq/kg, the area right of the limit represents the amount of food that cannot be sold to consumers. Countermeasure policies at the stage of production enable the selection of cultivars with reference to concentrations of radioactive contamination, with the intention of creating conditions where transfer to food is limited to the greatest extent possible. For example, the adoption of an agricultural countermeasure to reduce a crop's radioactive contamination from 5 to 1 Bq/kg is to take action to preventatively control the transfer of radioactive materials to food products. As depicted in Figure 6.1, such production-stage countermeasures will result in a shift in the distribution of food to the left, represented by the dashed curve.

Production methods in agriculture, livestock, forest, and fisheries are all different and, accordingly, radiation countermeasures must also differ for each type of

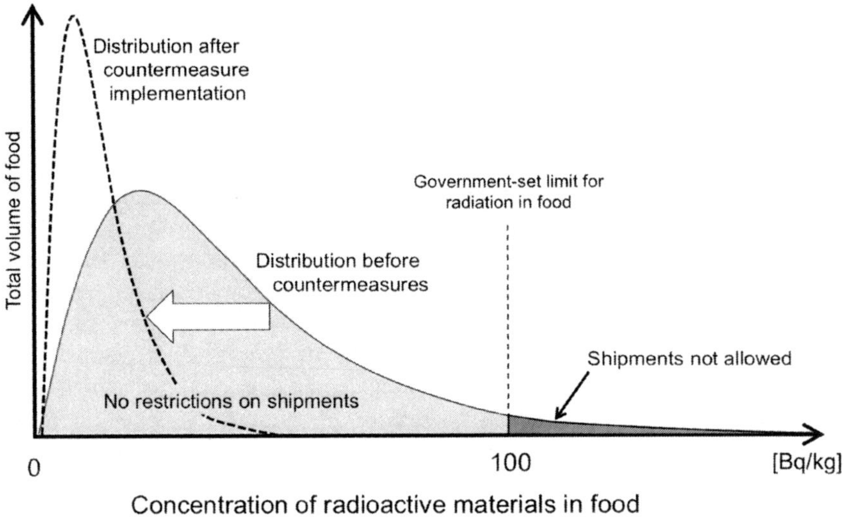

Figure 6.1 Model distributions for the contamination of radioactive materials in food.

Source: Author.

food production. For agricultural products and livestock, the many decisions pertaining to where and what is cultivated, what kind of soil treatments are conducted and fertilizer used, and what kind of feed is given to livestock are all illustrative of the fact there is ample leeway for controlling radiation through alterations in the conditions of production. In other words, if agricultural and livestock production are conducted using uncontaminated soil, water, and feed, then the transfer of radioactive materials can be minimized. Additionally, in the case of agricultural crops, since the transfer of radioactive materials to crops is influenced by the chemistry and physical state of the soil, there is ample room for controlling transfer through soil treatment and fertilizer application. This is to say that, in the case of "cultivated products," policies can be applied from the stage of production. Today, few agricultural products from Fukushima exceed the 100 Bq/kg limit for radiation. In fact, most products are found to be under a few Bq/kg. To make policies regarding these products more thorough, it is best to implement them at the stage of production rather than to strengthen food testing.

In contrast to cultivated food products, it is difficult to implement policies at the stage of production for wild forest and fisheries products gathered from the bounty of nature. Accordingly, the only feasible policies at the stage of production are to identify areas where forestry and fisheries products below the limits for radiation can be produced, and to regulate areas where harvesting can be conducted. In short, in the case of foods that are procured through hunting, fishing, and gathering, food inspection becomes paramount. Indeed, there are many cases in which mushrooms and wild edibles, wild animals such as boar and bear, and demersal fish have been found to be over the 100 Bq/kg limit, and it is therefore highly important to emphasize radiation testing for such foods for the time being.

Mapping the distribution of radioactive materials

In order to devise appropriate production-stage policies to control radioactive contamination of food, it is essential to accurately measure the spatial distribution of radioactive materials. The radiation-monitoring system implemented by the national government is far from sufficient in terms of its spatial resolution. To supplement the existing monitoring scheme, the agricultural cooperative JA Shin-Fukushima has used the Nal Spectrometer developed by the ATOMTEX Company of Belarus (AT6101DR) to pursue radiation measurement for every individual paddy field and orchard in Fukushima City (Figure 6.2). As of the end of 2014, radiation measurements for roughly 24,000 paddy fields and around 10,000 orchard fields had been completed. This measurement of farmland has been conducted through a collaborative joint effort between agricultural cooperative staff and the staff of the Japanese Consumer Co-operative Union gathered from around the country. Through this joint effort, the pairing of the different interests of producer and consumer has secured the objectivity of radiation-measurement activities.

The device noted above detects gamma rays emitted from the soil, and can quantify the concentration (Bq/kg) and deposition (Bq/m^2) of cesium 134 and 137 and potassium 40 in soil. It is also equipped with GPS and records latitude, longitude, and altitude; it can thus be used to visualize data over air photos on Google Earth. The most remarkable feature of this device is that it can be used without collecting soil and can take a measurement within the very short span of

Figure 6.2 Tools of soil-radiation measurement.

Source: Author.

two minutes, thus allowing for timely analysis of the distribution of radioactive materials over a wide area.

Fukushima University has worked to uniformly compile the results of this large effort to measure agricultural fields, and has developed software for extracting and listing the data as well as making the data transferable to GIS and databases. As a result, not only have the results been visualized on Google Earth, but it has also become possible to: 1) link individual field data (owner information, cultivation history, soil chemical composition, etc.) with measurement-result data; 2) conduct various analyses based on GIS and statistical analysis software; 3) represent measurement results in various map forms; and 4) return measurement results to individual landowners. After making the results of measurements known to landowners, there are plans to link this data with the results of Total-Volume-All-Bag-Testing (discussed below), in order to provide farming guidance to landowners.

The mechanism behind the absorption of radioactive cesium by plants

Following the nuclear disaster, radioactive materials were emitted into the atmosphere and later deposited, resulting in the contamination of soil and crops. There are two ways in which radioactive materials can contaminate crops (Kawatsu, Ohse, and Kitayama 2016). The first occurs when radioactive materials (primarily cesium) deposited from the atmosphere attach directly to crop surfaces and remain there, or are absorbed into a crop through “foliar absorption.” The second occurs when radioactive materials contained in soil or water are absorbed by crops through “root absorption.”

Radioactive materials that adhere to crop surfaces can be particularly damaging for long-lived plants such as fruit trees. In Fukushima Prefecture, where fruit-tree cultivation is a leading agricultural activity, decontamination practices deploying high-pressure sprayers to remove radioactive materials from the bark of trees were widely implemented in the winter of 2012. The Fukushima nuclear disaster was enormous, but contamination of crops was somewhat mitigated by the fact that the accident occurred in March. Most crops, and most importantly rice, were not under cultivation at the time of the accident and most fruit trees had not yet put out leaves, thus avoiding a large degree of surface attachment and absorption. Indeed, the fact that so little contamination of food resulted from such a large accident was the silver lining to an otherwise very dark set of clouds. If a nuclear disaster had occurred after the beginning of the main farming season, in May, the leaves and stems of growing crops would have absorbed cesium and an enormous degree of food contamination would have occurred.

It came as a shock, therefore, when two years after the accident, in 2013, radioactive contamination of agricultural crops—apparently by atmospheric deposition—was detected. It turned out that most of the rice over the 100 Bq/kg limit came from Minami-Soma City. An emergent explanation for this occurrence was that when wreckage was removed from the Fukushima Daiichi plant in August 2013, radioactive cesium from the area was scattered in the direction of Minami-Soma, leading to the deposit of cesium on rice and foliar absorption or contamination of the water supply. The precise cause and mechanism are still under debate, and a research team at Fukushima University is presently conducting experimental research to determine the cause of this contamination. It seems safe to say, nevertheless, that except through incidents such as this, serious food contamination—such as that seen immediately after the disaster—is unlikely to reoccur. What is urgently needed, therefore, is to prevent the re-dispersion of radioactive materials from Fukushima Daiichi and to strengthen atmospheric monitoring systems, which enable swift response—including inspection of food and restrictions on product distribution—in preparation for similar incidents.

With regard to root absorption, the rate of plants' absorption of radioactive cesium is determined by such factors as crop type and the physical state and chemistry of soil. In paddy-field agriculture, the rate of absorption is also influenced by the concentration of radioactive cesium and inorganic ions in water in fields. The transfer coefficient, which is given by the concentration of radiation in crops (Bq/kg) divided by the concentration of radiation in soil (Bq/kg), is useful as a general guideline for determining plants' rate of cesium absorption from soil.

The specific coefficient value is determined for each part of a crop (e.g. root, stem, fruit, etc.) and is influenced by the conditions of production (Figure 6.3). Most soils in Japan are rich in clay minerals, which absorb cesium at a relatively high rate (in comparison with soils in Belarus or Ukraine, for example), tending to suppress the transfer of cesium to crops. If the concentration of radioactive cesium in soils is known, then the transfer coefficient can be used to predict transfer to crops to some extent. At sites with high levels of contamination, crops with low transfer coefficients can be cultivated to produce relatively uncontaminated products, and such knowledge is useful in formulating plans for cultivation. It should be noted, however, that it takes some time for cesium to attach to clay minerals. Therefore there may have been a comparatively rapid progression of cesium transfer immediately after the accident.

Early attempts to determine the cause of tainted rice

The transfer coefficient of rice is quite low, at around 0.001. Based on this low transfer coefficient, in 2011 it was expected that rice produced in soils containing under 5,000 Bq/kg would not be over the temporary limit of 500 Bq/kg for radiation in food. Accordingly, in areas where evacuation designations had not been stipulated, rice planting was permitted. However, in the autumn of 2011, rice containing over 1,000 Bq/kg was identified, giving producers and consumers quite a shock.

Inspection of paddy fields where cesium absorption by rice plants was prominent failed to reveal any correlation between degree of soil contamination and cesium absorption. Furthermore, the apparent transfer coefficient, between 0.1 and 0.4, was extremely high. It was observed that paddy fields with low amounts of exchangeable potassium in soil tended to exhibit prominent cesium absorption. Furthermore, the paddy fields responsible for producing rice that was over the temporary limit for radiation in food were found to be distinctive landscapes: mountainous valley recesses surrounded by forests.

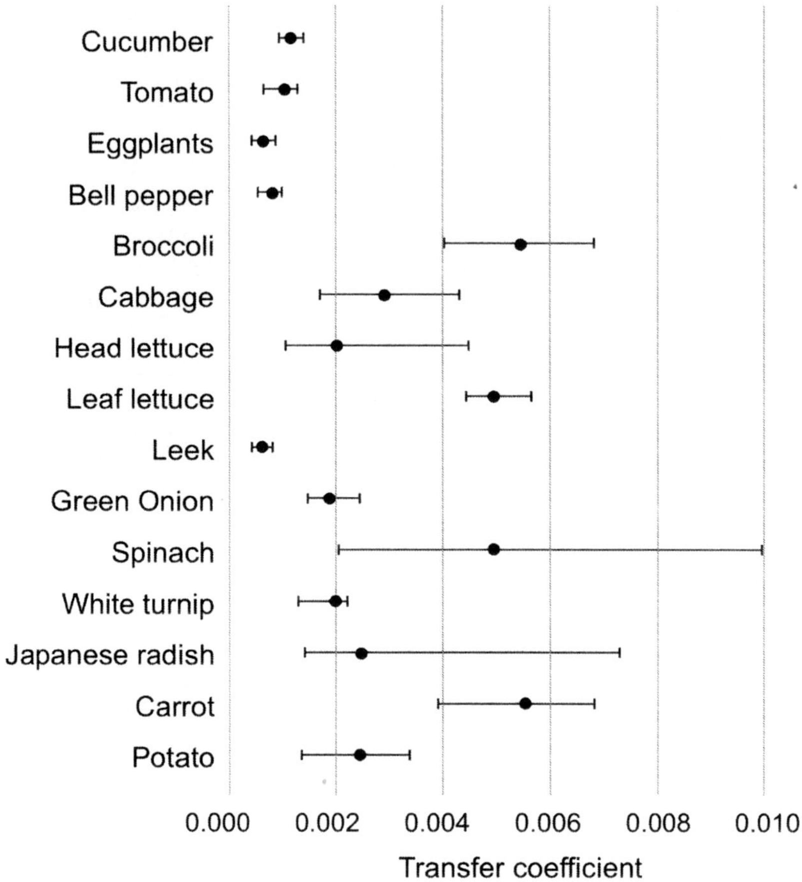

Figure 6.3 Transfer coefficients for different crops.

Note: Created by the author based on National Agriculture and Food Research Organization (2001b).

These observations led researchers to suspect two possible factors. The first was the role of contaminated water, rather than just soil, in influencing the rate of cesium transfer. The second was the effect of exchangeable potassium in soil. In terms of the first factor, experiments by Nemoto and Abe (2013) identified that absorption would be particularly prominent in plants cultivated in water containing cesium.[2] Based on such cultivation experiments and on circumstantial evidence about the environment of rice paddies producing contaminated rice, it was inferred that cesium absorption occurred outside soil, and the following, more specific hypothesis was constructed: increased cesium transfer to rice is caused by cesium contamination of the water supply for rice paddies, which is particularly pronounced in deep mountain recesses surrounded by forests.

The effects of exchangeable potassium in soil—the second factor—are depicted in Figure 6.4, which illustrates the relation between the amount of exchangeable potassium and absorption of cesium by rice. The horizontal axis of the graph represents the concentration of exchangeable potassium in soil while the vertical axis represents the concentration of radioactive cesium in soil. Plotted in this figure are data for rice from areas where contaminated rice was identified. As can be seen, when exchangeable potassium in soil is depleted, cesium absorption by rice is prominent, and when exchangeable potassium is retained, the tendency is for cesium absorption to be limited. Accordingly, it has been concluded that when fertilizers such as potassium chloride and potassium silicate are applied to increase exchangeable potassium in soil to over 25mg/100g, cesium absorption by rice can be significantly regulated (National Agriculture and Food Research Organization 2011a). This knowledge provided the basis for the wide recommendation within Fukushima Prefecture, starting in 2012, to use fertilizers to increase exchangeable potassium levels in soil when cultivating rice.

However, even when relatively sufficient exchangeable potassium levels of around 15mg/100g are retained, a type of "outlying" paddy field where cesium absorption remains high has also been identified (four points circled in Figure 6.4). The cause of these "outlying" paddy fields remained unknown in 2011, and was only elucidated after 2012 through ongoing experimental cultivation.

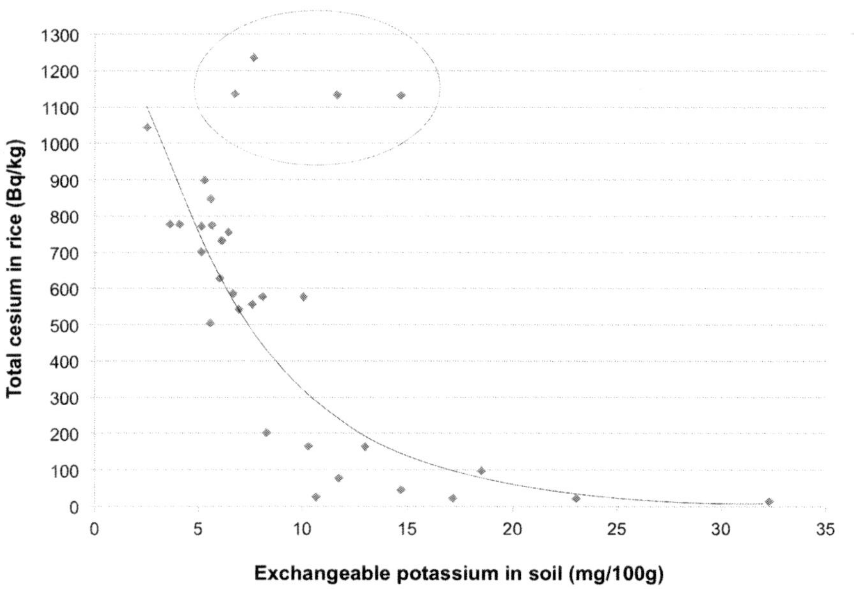

Figure 6.4 Effects of exchangeable potassium in soil on the level of cesium in rice.

Note: Created by the author based on Fukushima Prefecture and the Ministry of Agriculture, Forestry and Fisheries (2011). The circle was added by the author.

Experimental rice cultivation

In 2012, areas in which rice over the temporary limit for radiation in food had been identified in 2011 were designated “no planting” zones for rice. In addition, in areas that were not over the temporary limit, but where rice over 100 Bq/kg was identified, cultivation was conducted only under the following four conditions: 1) provision of a management ledger to each individual producer; 2) lowering of air-dose levels through deep plowing; 3) determination by local governments of the amount of potassium fertilizer and zeolite to be applied to fields; and 4) radiation testing for all rice.

Along with these practices, experimental cultivation has been carried out in the “no planting” zones. Experimental cultivation has three main significances. First, in order to lift the ban on planting, it must be determined whether the use of cesium-reduction materials, such as potassium chloride, will enable the production of rice under 100 Bq/kg. Results based on experimental cultivation at 396 sites in the “no planting” zones in Fukushima found only one case over the 100 Bq/kg limit. Accordingly, in 2013 the ban on planting was lifted for most of the “no planting” zones. The second reason for conducting experimental cultivation was to investigate the effects of soil and fertilizer conditions. The third significance of experimental cultivation was to evaluate the influence of water, particularly in relation to topography. As suggested by these different aims, experimental cultivation was conducted with reference to several different considerations. In collaboration with researchers at the University of Tokyo and Tokyo University of Agriculture, researchers at Fukushima University used 55 paddy-field sites in the Oguni District of Date City to conduct experiments. Without using materials such as potassium fertilizer or zeolite, cultivation was first conducted per usual methods to test the “paddy-field environment as it exists,” then potassium silicate was applied to some sections to test its effects.

Based on the results of experimental cultivation, the report published by Fukushima Prefecture and the Ministry of Agriculture, Forestry and Fisheries (2013) concluded that: 1) a correlation between concentrations of radioactive cesium in soil and radioactive cesium in rice could not be found; 2) potassium in soil counteracts absorption by crops and thus suppresses the transfer of cesium, and when potassium fertilizer is used to raise the amount of exchangeable potassium in soil above the target of 25mg/100g, the concentration of radioactive cesium in rice is greatly reduced. In addition, with regard to the effects of water supplies contaminated by radioactive cesium, the same report states that in contrast to plants’ capacity to directly absorb dissolved cesium, it is difficult for plants to directly absorb suspended cesium, and it can be thought that transfer is generally low. The report also suggests, drawing on the results of water-quality testing of irrigation canals and retention ponds conducted in Fukushima Prefecture, that the influence of water is limited in general.

However, in the experimental cultivation conducted in the Oguni District of Date City, high levels of cesium absorption were witnessed even in sites with exchangeable potassium in soil at the low levels of between 12 and 15mg/100g, similar to the “outlying” fields discovered in 2011 (Oguni District Experimental

Cultivation Support Group 2012). In the experimental cultivation study, these "outlying" fields were considered to have been caused primarily by cesium contamination of water sources. Accordingly, this study highlighted that: 1) water drawn into these "outlying" fields is from water sources containing, in some parts, 4 Bq/liter of radioactive cesium (mostly suspended cesium); 2) when comparing the concentration of radioactive cesium in leaves and stalks in mid-July and mid-August, "normal" fields tend to show higher values in July while "outlying" fields occasionally show prominent absorption in August.

The mechanisms behind rice plants' absorption of radioactive cesium remain inadequately elucidated. Nevertheless, we are quite certain at this point that the transfer of radioactive cesium in soil is generally low, but that there are two mitigating factors in existence that interact on the ground in agricultural fields in a complex fashion: 1) the potential for deficiencies in exchangeable potassium in soil to promote cesium absorption; 2) the potential for cesium-contaminated water sources to promote cesium absorption. We have not yet confirmed the causes of high concentrations of radioactive cesium in water sources. One possible explanation is that leaves with cesium adhered to them often accumulate in retaining ponds and other water sources, and if that cesium decomposes and becomes dissolved or suspended cesium, thus contaminating water sources, this could be the source of the transfer of cesium to rice plants. What this all suggests is that the problem of contaminated rice must be investigated from the perspective of the circulation of cesium through forest and agricultural environments by water.

Total-Volume-All-Bag-Testing for rice

One of the last points at which the contamination of agricultural food is controlled is the testing of harvested food before it enters into distribution channels. Here I discuss the procedure of food testing, focusing on rice. In the fall of 2012, Fukushima Prefecture began to conduct Total-Volume-All-Bag-Testing for all rice produced in the prefecture. In this testing procedure, a 30 kg bag of rice is placed on a belt conveyor for around 15 seconds and the radiation level is measured. This testing method has a lower detection limit of 10 Bq/kg. Table 6.1 depicts the results of Total-Volume-All-Bag-Testing between 2012 and 2015.

In 2012, bags of rice containing over 50 Bq/kg were found not only in the prominently contaminated areas of Fukushima City, Nihonmatsu City, and Date City (e.g. soils containing roughly 5,000 Bq/kg), but also in relatively lightly contaminated areas (e.g. 2,000 Bq/kg). In these areas of relatively low soil contamination, contamination of rice was caused by cultivation in paddy fields where environmental conditions promoted absorption, and where local governments did not recommend policies for applying potassium fertilizers. However, in the rice paddies containing 5,000 Bq/kg fields under the jurisdiction of "JA Date Mirai," no rice was found to be over the limit of 100 Bq/kg. This was due to the fact that the agricultural cooperative, in collaboration with producers, applied 200 kg of potassium chloride per 1000 m^2 of field. In 2014, (only) two bags of rice over the 100 Bq/kg limit were identified in all of Fukushima. The cause of this phenomenon was later determined to be failure to apply potassium chloride, which

Table 6.1 Results of Total-Volume-All-Bag Testing between 2012 and 2015

	2012	*2013*	*2014*	*2015*
Under 25 (Bq/kg)	10,323,674 (99.78%)	10,999,222 (99.93%)	11,012,562 (99.98%)	10,231,706 (99.99%)
25–50	20,357 (0.2%)	6,484 (0.06%)	1,910 (0.02%)	610 (0.006%)
50–75	1,678 (0.016%)	493 (0.0045%)	12 (0.0001%)	17 (0.0002%)
75–100	389 (0.0038%)	323 (0.003%)	2 (0.00002%)	1 (0.00001%)
Over 100	71 (0.0007%)	28 (0.0003%)	2 (0.00002%)	0 (0%)
Total	10,346,169 (100%)	11,006,550 (100%)	11,014,488 (100%)	10,232,334 (100%)

Unit: Count of 30 kg rice bags.

Source: Fukushima no Megumi Anzen Taisaku Kyogikai (https://fukumegu.org/ok/kome/; accessed July 8, 2015).

created conditions that facilitated cesium absorption. If potassium chloride had been applied, then rice under the limits would have been produced. What these episodes indicate is first, that we must take measures for the most at-risk fields, and second, the importance of organizational management at the levels of government and agricultural cooperatives accurately implementing reduction measures, without leaving the responsibility to individual producers.

Developing sustainable countermeasures

In the five years that have passed since the nuclear disaster, scientific knowledge has been accumulated and many organizational mechanisms developed. The task for the future is to transition from the trials and responses of the emergency period immediately after the disaster toward sustainable permanent policies.

As emphasized at the outset of the chapter, policies at the stage of production and postharvest food testing are inseparable. While this chapter has introduced the trends and outcomes of experimental cultivation, radioactive material distribution mapping, and Total-Volume-All-Bag-Testing as separate endeavors, it is only by linking these policies together in a mutually reinforcing policy system that greater results can be achieved. For example, if fields at high risk of cesium absorption are identified through Total-Volume-All-Bag-Testing, then soil analysis can be employed to determine whether cesium absorption was caused by deficiencies in exchangeable potassium or contaminated water supplies, and the state of risk can be evaluated. Subsequently, if a record is created for each individual agricultural plot, it should be possible to provide farming guidance to individual producers that is reflective of the particular conditions of their environment. Furthermore, by accumulating such records and knowledge, it becomes possible to explicate cesium circulation in forest and agricultural environments and the mechanism behind the absorption of cesium by rice plants. In this manner, the results of Total-Volume-All-Bag-Testing provide knowledge that is extremely important to the formulation of policies at the stage of production.

Conversely, if maps of radioactive material distribution are created and knowledge of the mechanism behind the transfer of radioactive materials to plants is gained, then the accuracy and validity of Total-Volume-All-Bag-Testing can be verified and the reliability of food testing increased. Essentially, the impetus to conduct Total-Volume-All-Bag-Testing was brought about by the fact that there was no knowledge of the area in which bags of rice over the 100 Bq/kg limit were being produced. For example, if paddy fields in a certain locality do not produce any detected rice for a few years, and if the production conditions are not conductive for cesium absorption, then in the future, upon obtaining social consensus, these areas can be removed from Total-Volume-All-Bag-Testing. If this materializes, it can be linked to reducing the tremendous labor, cost, and time that stems from Total-Volume-All-Bag-Testing. Additionally, focusing only on fields with high risk will make it feasible to conduct even more rigorous analysis.

Additionally, if paddy fields in need of potassium chloride and other reduction materials can be distinguished from fields not needing such materials, then the current system of uniform application decided at the local-government level can be converted into a field-by-field system reflective of individual productive conditions and farming guidance. If this materializes, it will become possible to accurately identify paddy fields where measures must be taken, "policy leakage" can be eliminated, and eventually this will reduce the risk of stigma to producers. It would also reduce the labor and costs associated with reduction materials and increase the sustainability of policies.

In order to secure a safe and reliable food supply, as outlined above, it is important not only to gain scientific knowledge such as an understanding of the transfer mechanism of radioactive materials to crops and the environment, but also to bring these results back to the local scene. To reliably enact countermeasures, it is essential to identify the radioactive contamination and productive conditions of each agricultural field and to construct a system capable of providing detailed farming guidance to each individual producer.

Acknowledgements

This work was supported by JSPS KAKENHI Grant Number 25220403.

Notes

1 For general explanation of the meanings of units of radiation and radioactivity, see Yamakawa and Yamamoto (2016, 8–11).

2 Cultivation in water containing 0.1 Bq/liter resulted in rice plants containing 76 Bq/kg; cultivation in water containing 1 Bq/liter resulted in rice plants containing 560 Bq/kg; and cultivation in water containing 10 Bq/liter resulted in rice plants containing 5,600 Bq/kg.

References

Fukushima Prefecture and Ministry of Agriculture, Forestry and Fisheries. 2011. *Zantei Kiseichi wo Chosetsushita Hoshasei Sesiumu wo fukumu Kome ga Seisan sareta Yoin no Kaiseki* [Analysis of Causes of the Production of Rice Containing Radioactive Cesium above the Temporary-set Standard Levels]. Accessed January 20, 2014. www.pref.fukushima.lg.jp/uploaded/attachment/10425.pdf. [In Japanese].

Fukushima Prefecture and Ministry of Agriculture, Forestry and Fisheries. 2013. *On The Causes of High Concentrations of Radioactive Cesium in Rice And Countermeasures: Research into Causes and Experimental Cultivation.* Accessed January 1, 2016. www.maff.go.jp/j/kanbo/joho/saigai/pdf/youin_kome2.pdf. [In Japanese]

Kawatsu, Kencho, Kenji Ohse, and Kyo Kitayama. 2016. "Outline of an Invisible Disaster: Physio-Spatial Processes and the Diffusion and Deposition of Radioactive Materials from the Fukushima Nuclear Accident." In *Unravelling the Fukushima Disaster*, edited by Mitsuo Yamakawa and Daisaku Yamamoto, pp. 22–37. London: Routledge.

Kimura, Aya H., and Yohei Katano. 2014. "Farming after the Fukushima Accident: A Feminist Political Ecology Analysis of Organic Agriculture." *Journal of Rural Studies* 34: 108–116.

National Agriculture and Food Research Organization. 2011a. *Genmai no Hoshasei Sesium Teigen no Tameno Kari Shiyo* [The Use of Potassium to Counter the Absorption of Cesium by Rice]. Accessed June 30, 2016. www.naro.affrc.go.jp/publicity_report/press/laboratory/narc/027913.html. [In Japanese]

National Agriculture and Food Research Organization. 2011b. *Kakushu Kasaku Yasai eno Dojochu no Hoshasei Sesiumu no Iko Keisu* [Transfer Coefficients of Cesium Within Soils into Various Summer Vegetables]. Accessed June 30, 2016. www.naro.affrc.go.jp/project/results/laboratory/tarc/2011/a00a0_01_77.html.

Nemoto, Keisuke and Jun Abe. 2013. "Radiocesium Absorption by Rice in Paddy Field Ecosystems." In *Agricultural Implications of the Fukushima Nuclear Accident*, edited by Tomoko M. Nakanishi and Keitaro Tanoi, pp. 19–28. Springer Open. Accessed June 29, 2016. http://link.springer.com/book/10.1007%2F978-4-431-54328-2.

Oguni District Experimental Cultivation Support Group. 2012. *Oguni Chiku Niokeru Ine no Shikensaibai ni Tsuite* [On the Experimental Cultivation of Rice in the Oguni District]. Accessed June 29, 2016. www.a.utokyo.ac.jp/rpjt/event/2012120805-2.pdf. [In Japanese]

Yamakawa, Mitsuo and Daisaku Yamamoto (eds). 2016. *Unravelling the Fukushima Disaster*. London: Routledge.

7 Resilience of local food systems to the Fukushima nuclear disaster

A case study of the Fukushima Soybean Project

Takashi Norito

Introduction

The triple disaster of the Great East Japan earthquake and tsunami and the subsequent accident at TEPCO's Fukushima Daiichi Nuclear Power Plant brought unexpected and unprecedented challenges to households, local communities, agriculture, and industries in the afflicted regions. This chapter concerns the recovery of food production in Fukushima. In particular, I focus on the recovery of *food systems*, interconnected processes of food production, processing, distribution, and consumption (Niiyama 2001; Takahashi and Saito 2002; Tansey and Worsley 1995), rather than simply examining the recovery of agriculture per se.[1] In this view, the historic disaster can be seen as a major disturbance to the workings of food systems.

A food system is typically defined by a set of connected processes and activities around a particular food commodity, such as food production and manufacturing, wholesaling, food services, and retailing. It also encompasses various socio-political actors including farmers, workers, consumers, local and national governments, and other related institutions. In this sense, food systems are necessarily social systems. Furthermore, each food system has its own geography, whose extent varies widely from global, involving multinational corporations, to local, based on the model of community-supported agriculture (CSA). Although there is little doubt that geographically extensive, global food systems are growing in significance (Fold and Pritchard 2005; Whatmore 1995), local food systems remain crucial especially to revitalize regional economies, to strengthen local governance and autonomy, and ultimately to enhance the quality of civil society. Finally, each food system is characterized by its own historical trajectory. In rural areas of Japan, there have been an increasing number of active local movements to connect food producers with consumers supporting them through buying local foods, aiming to develop more sustainable regional economies. For example, long before the nuclear disaster, cooperative and private sectors had been working to build a variety of local food systems in Fukushima. These efforts include farmers' markets, CSA based on organic produce, local sourcing of ingredients by local food manufacturers, and collaboration between different types of local cooperatives.

The radioactive contamination and its associated stigma (*fuhyo higai* or "reputational damage") critically harmed local food systems in Fukushima Prefecture. However, it is important to note that many local food systems, though damaged, did not completely break down, and that actors including farmers, manufacturers, and cooperatives are striving to overcome the challenges raised by the nuclear disaster. The Fukushima Soybean Project (FSP) is a case in point, and is the focus of this chapter. The project started in 1998, to locally process and consume soybeans grown in Fukushima through a partnership among agricultural cooperatives responsible for production, private firms responsible for processing, and consumers' cooperatives responsible for distribution and consumption. The project has been successful and gained much support from participating individuals and organizations. The nuclear disaster has presented many challenges, but to date the project continues to operate. By articulating these challenges and efforts to overcome them, this chapter offers empirical insights into the resilience and rebuilding of local food systems following significant external disturbances.

This chapter is organized as follows. The next section briefly reviews the trends of local food systems in Japan. In particular, it emphasizes the social significance of local food systems, which help local communities to regain food sovereignty and restore a strong local identity, and their economic significance, in promoting the local agriculture and food-processing industries by fostering their linkages. The section following that one presents an overview of how local food systems in Fukushima have been affected by the nuclear disaster. This is followed by an overview of the FSP, established in the late 1990s, which is managed by a collaboration between agricultural and consumers' cooperatives. This collaboration illustrates effective mechanisms for building trust that have helped to cope with the nuclear disaster. The next section also examines the FSP's response to the challenges brought about by the nuclear disaster, especially shortages of local ingredients due to actual radioactive contamination, and consumers' anxieties about its risks. The ways in which the FSP coped with these challenges may provide useful lessons for other food systems affected by nuclear accidents. The final section summarizes the findings and discusses their theoretical and practical implications.

Local food systems: conceptual origins and trajectories in Japan

Interest in local food systems has grown over the past decades for two important reasons: first, as they are alternative systems of food distribution, and second, as they are a means of regional economic revitalization. In specific regional and national contexts, one of the reasons may overshadow the other. In the United States, for example, such systems tend to be framed primarily as alternative means of food distribution that counter the mass industrialization and globalization of food systems (Lyson 2004). Lyson observes these movements based on the localization of agriculture and food production not only in the United States, but also in other industrialized countries. He calls such a resurgence of community-based agriculture and food production "civic agriculture." Henderson and Van En

(2007) similarly examine and promote various mechanisms, such as CSA in the United States, that link farmers who emphasize environmental quality and food safety with consumers who support such producers through their consumption. In a similar vein, the idea of "short food-supply chains" captures the reduction of socio-spatial distance between food production and consumption (e.g. Morris and Buller 2003; Renting *et al.* 2003).

In Japan, a similar movement occurring since the 1990s has been called *chisan-chisho*, which literally means "local production, local consumption" (Ito 2012). This movement has several predecessors. First, Japanese consumer cooperatives have been practicing the *sanchoku* ("direct delivery from producers") system, as a way to bypass wholesalers and retailers in transaction, since the 1970s. Second, the *teikei* ("partnership") system, developed as a part of organic agriculture movements also in the 1970s, facilitates direct contracts between consumers and farmers (similar to CSA). Nevertheless, as pointed out by Nishiyama and Kimura (2005) and Kimura and Nishiyama (2008), these earlier movements tended to focus primarily on the needs of individual consumers (for safe, fresh, and reasonably priced food), with those of farmers (for stable sales and higher profit margins) considered to a lesser extent. Unlike these earlier attempts to develop alternative food-distribution systems, the more recent *chisan-chisho* movement has a stronger interest in benefiting local communities in which farmers and other related actors work and live, implying a shift in emphasis from consumer movements to local-community movements. In the *chisan-chisho* movement, therefore, goals such as the inheritance of local food-and-farming culture through education at local schools, along with deepening ties between producers and consumers, are often emphasized.

The second motivation for promoting local food systems—the revitalization of regional economies—often reflects the growing concern over food self-sufficiency of the nation and distressed rural peripheries. On the policy front, these issues are often addressed by the formation of food-industry clusters to strengthen relations among related actors in food production, processing, distribution, and marketing. The policies have been promoted strongly by governments not only in Japan but also in other East Asian countries, such as South Korea and Taiwan, since the early 2000s (Chung 2009; Hwang 2008). It is important nevertheless to distinguish two different theories of local economic growth. The export-base theory, originally proposed by North (1955), essentially argues that regional economies grow by exporting goods and services. The theory differentiates "basic" industries from "non-basic" ones in a region. A basic industry sells goods and services outside the region to bring in income, while a non-basic industry supplies goods and services within the region. Here the growth of non-basic industries is ultimately attributed to the growth of basic industries (i.e. exports). Many of the contemporary regional economic-growth theories, including industrial-cluster theory (Porter 1998), still share this fundamental assumption: export drives regional economic growth.

Markusen (2007) takes up and criticizes this assumption and proposes a consumption-base theory of development by arguing that expanding local consumption to substitute for previously imported goods and services, such as cultural and leisure consumption, can be as important a source of local economic growth. Consumption-base theory echoes the work of Okada (2005), which argues, with numerous empirical illustrations from local initiatives in Japan, that regional economic actors reinvesting within their own region can critically facilitate the circulation of income, expand local consumption, increase wages and profits, and ultimately revitalize regional economies. While I do not deny the role of export incomes in regional growth, the significance of local consumption and reinvestment should not be overlooked. In the context of local food systems, what this means is that a deep integration of local agriculture, food processing and manufacturing, related industries (tourism, education, health, etc.), and consumers, combined with appropriate institutional support from governments, can have positive effects on regional economies through local multiplier effects and job creation (Figure 7.1).

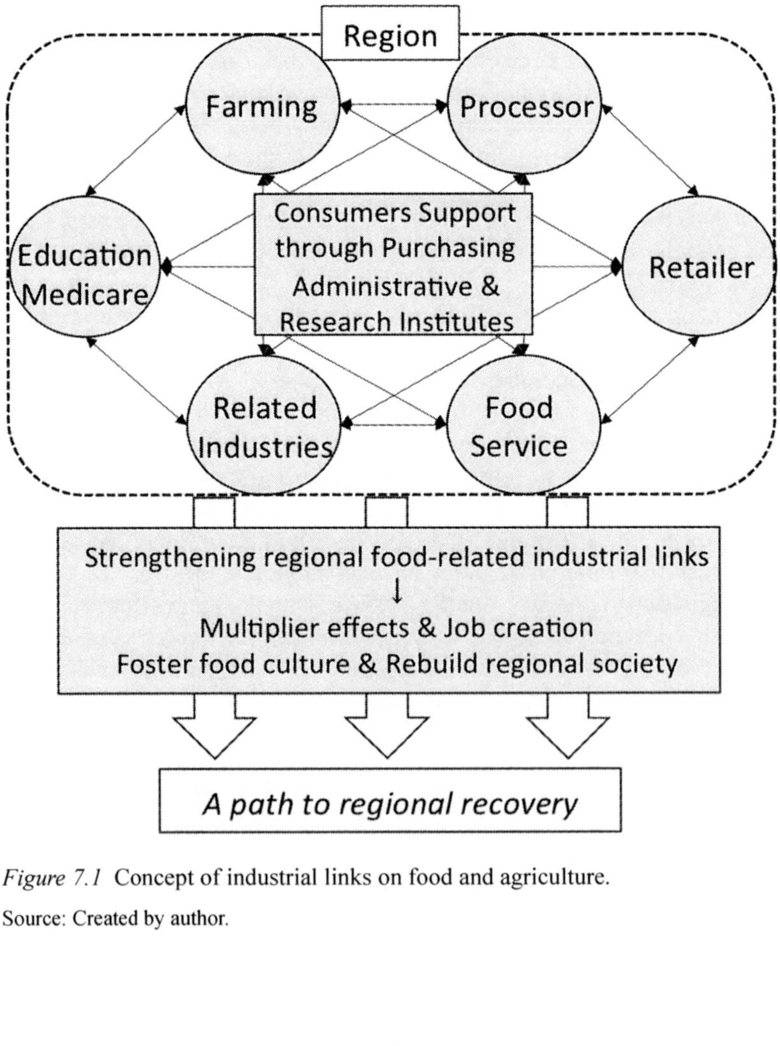

Figure 7.1 Concept of industrial links on food and agriculture.

Source: Created by author.

In summary, local food systems can be seen as means of both social movements and regional economic development, and I believe that local food systems can play a key role in the rebuilding of local communities in Fukushima afflicted by the nuclear disaster and more broadly in Japan's future regional policies. In particular, it is worth investigating how local food systems can be resilient to such a large, unexpected "disturbance" as a nuclear disaster. In this way, the analysis presented in this chapter has important implications for the growing literature on regional and local resilience (Yamamoto 2011).

Effects of the nuclear disaster on local food systems

Fukushima Prefecture, due to its mild climate and proximity to the large metropolitan market of Tokyo, has developed a diverse array of agriculture ranging from rice—the main crop—to fruits, vegetables, dairy, and livestock. Thanks to the diversity of produce, food-processing industries have also flourished in the prefecture. Farmers and processing firms often organize producers' unions around specific food products, such as *sake* (rice wine), Japanese-style pickles, wheat and buckwheat noodles, dairy products, soy sauce, *miso* (fermented soybean paste), *tofu*, and *natto* (fermented soybeans). Most of the processing firms are small and medium-sized enterprises (SMEs), some of which strategically use local ingredients and sell locally branded products to local retailers and consumers in order to differentiate their products from those of larger firms that target national markets.

In retailing, there has been a nationwide trend of increasing dominance of large, chain supermarkets, headquartered in metropolitan areas. This has also been the case in Fukushima—a good many local supermarkets and grocery stores have closed in the past 20 years—but some locally based supermarkets and cooperatives still remain in business. These stores typically make a strong effort to differentiate themselves from large chain stores by promoting locally produced and processed food. In this sense, these locally based stores have been also important actors in local food systems in Fukushima.

It goes without saying that the Great East Japan Earthquake Disaster had a profound impact on local food systems in Fukushima (Figure 7.2). First, the point of production was damaged by the earthquake, tsunami, and radioactive contamination (Koyama 2013). In particular, radioactive contamination inflicted the most significant damage and enduring uncertainties on farmers. Radioactive materials that exceeded the legal limit were detected in some rice and vegetables that were harvested after the nuclear explosions. The government put restrictions on the shipment of these food items and on the cultivation of crops in the fields in which high levels of radiation were detected. Meanwhile, in order to restore food safety, the government implemented, in haste, radioactive contamination-abatement measures in farmlands and radioactivity-monitoring systems for agricultural produce. Despite these measures, some Fukushima residents—particularly mothers with small children, who feared internal exposure to radiation—stayed away from locally grown produce. In response to their requests, some retailers and supermarkets in the prefecture stopped placing locally grown food on their shelves, which led many local food-processing firms to switch to ingredient suppliers who were overseas or outside the prefecture. This chain reaction damaged a variety of local food systems in the prefecture.

Figure 7.3 exhibits the trends in agricultural output values and food-manufacturing shipment values in Fukushima Prefecture. Although agricultural output decreased to 180 billion yen (approximately US$1.5 billion) in 2011, from

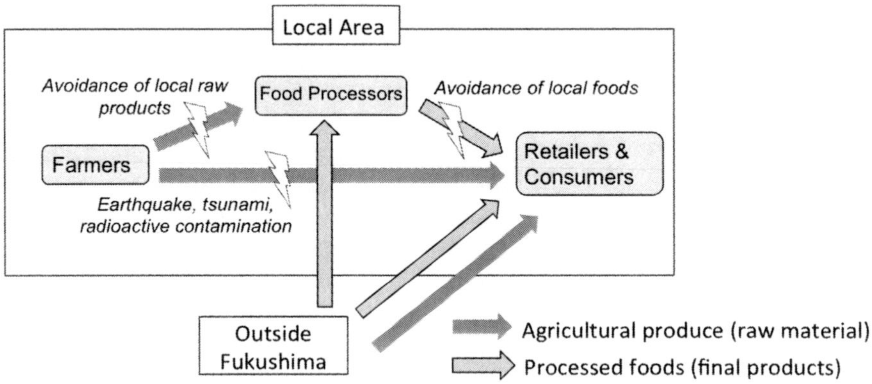

Figure 7.2 Local food systems damaged by the nuclear disaster.

Source: Created by author.

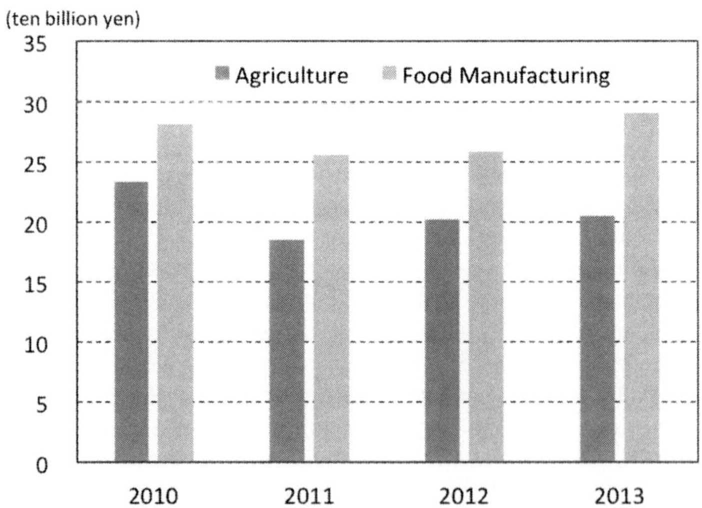

Figure 7.3 Trends of agricultural output and food-manufacturing production and shipment values in Fukushima Prefecture.

Source: Ministry of Agriculture, Forestry and Fisheries of Japan "Productive agriculture income statistics" and Ministry of Economy, Trade and Industry "Industrial statistics".

230 billion yen (US$1.9 billion) before the disaster, it recovered to 200 billion yen (US$1.68 billion) in 2012. Meanwhile, shipment values for the food-manufacturing industry plunged in 2011, though they had returned to pre-disaster levels by 2013. These data may suggest recovery on the part of agriculture and food manufacturing from the damage incurred, but by themselves they do not necessarily indicate the recovery of local food systems—relations among regional agriculture, the food-manufacturing industry, and consumers (Figure 7.2).

According to an annual survey conducted by a consumer group, the number of those who avoid local foods in Fukushima is declining, but nevertheless 20 per cent of residents continued to avoid local foods as of 2014. Another survey conducted by a local industry group also revealed that many food-processing firms switched their ingredients to those from non-local sources after the disaster, and many continue to use non-local ingredients even today. In other words, local food systems in Fukushima Prefecture are recovering, but they are certainly not back to where they were prior to the disaster. Under such circumstances, the FSP is trying to overcome the effects of radioactive contamination and stigmatization from the nuclear disaster. The next section examines the challenges that the project faced after the disaster through a case study of the FSP.

The Fukushima Soybean Project (FSP) before the disaster

Background of the FSP launch

Soybeans are an indispensable part of Japanese food culture, and are the main ingredient of fermented seasonings such as *miso* and soy sauce and of common processed foods such as *tofu* and *natto*. Many people think of rice as the most representative element of the Japanese food culture, but it is no exaggeration to say that soybeans are an equally essential part of it. Unlike rice, however, post-war Japanese agricultural policies excluded soybeans from the list of protected items early on. Soybean imports were liberalized in 1961 and cheap foreign imports came to dominate the market, in sharp contrast to rice, which has long been placed at the center of Japanese agricultural protection policies. In 1960, 60 per cent of edible soybeans and 30 per cent of overall soybeans (including soybeans for oil) were supplied domestically, but these shares plunged to 20 per cent for edible soybeans and around 5 per cent for soybeans overall within a decade after the liberalization. Currently approximately 80 per cent of edible soybeans and over 90 per cent of overall soybeans (including those for animal feed) are imported.

Meanwhile, since the late 1990s, genetically modified (GM) soybeans have become popular worldwide. As in many other countries, the Japanese food regulations require GM food labeling, and many Japanese consumers feel some concerns over GM food. Consequently, calls to increase national self-sufficiency in soybeans began to emerge on the grounds of both food security and food safety. Interest in resurrecting local production and consumption of soybeans needs to be understood in this context, and it was precisely the motivation behind the FSP.

Despite the increasing interest in local production and consumption of soybean products, there were structural challenges that had to be overcome. Traditionally, Fukushima was one of the leading producers of rice. Under the Japanese government's rice acreage-reduction policy since the 1970s, there had been a growing push to cultivate wheat, soybeans, and vegetables in the place of rice, and soybean production was on the rise in Fukushima. However, as of the 1990s the volume was still far too small, because farmers were unable to earn sufficient income from soybeans due to competition from cheap imports. Furthermore, the government had strong control over the distribution of some crops, including soybeans and wheat, at the national level, and made local circulation of products difficult.[2] Consequently, despite the presence of a large number of small and medium-sized firms marketing soy-based products, most soybeans from Fukushima were being shipped outside the prefecture, while the local processors used soybeans from outside the prefecture or overseas. Furthermore, because a sizable amount of soy-based products from Fukushima were sold outside of the prefecture, Fukushima consumers ended up relying heavily on non-local products. This was the challenge taken up by the FSP.

Mechanisms of "buying local"

The FSP was initiated in 1998 as a collaborative effort among agricultural and consumers' cooperatives and local food manufacturers to promote the local processing and consumption of soybeans within Fukushima Prefecture (Figure 7.4). The main products include *tofu*, *natto*, soy sauce, and *miso*, all essential ingredients of Japanese home cooking. At the beginning of the project, the main challenge was to secure sufficient local demand for locally produced and processed soybeans at an appropriate price. Fukushima-grown soybeans fetched between

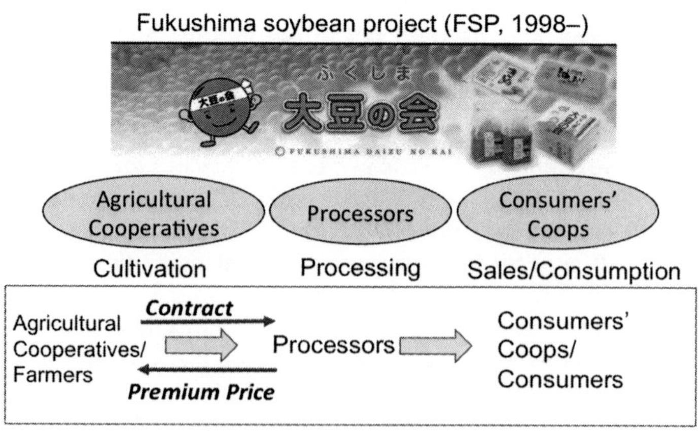

Figure 7.4 The Fukushima Soybean Project illustrated.

Source: Created from the FSP website (http://f-daizunokai.com/) and interviews.

5,000 and 13,000 yen (approximately US$40–110) per 60 kg on the wholesale market. Under the FSP, food-processing firms, such as *tofu* manufacturers, contracted with local agricultural cooperatives to purchase soybeans with a premium of 3,000 yen (US$ 25) per 60 kg on the market price. Consumers' cooperatives in turn collected members who would subscribe to regular purchases of locally grown and processed soybean products, which would be delivered to subscribers' homes (although some of the goods are also sold at cooperative stores and farmers' markets). The number of subscribers jumped from 1,700 to 10,000, which was the initial recruiting goal, within the first three years of the project. By securing a steady volume of demand, soybean-processing firms were able to keep the prices of their products only slightly higher than those of conventional products made from imported ingredients, rather than having to target extremely small, high-end niche markets. Nonetheless, for the processing firms, the business was still far from profitable, but their desire to contribute to the development of the region as locally based food companies led them to join and stay in the project.

Organizational factors to sustain growth

The FSP continued to expand over the decade after its launch (Figure 7.5). Sales of soybean products increased from 67 million yen (approximately US$564,000) in fiscal year 1999 to more than 250 million yen (US$2,100,000) in 2010. The substantial growth and success of the project can be attributed to collaborative, win–win relationships among agricultural cooperatives (producers), consumers' cooperatives (consumers), and manufacturers (processing firms), designed by the architects of the FSP. Particularly noteworthy is the collaboration between

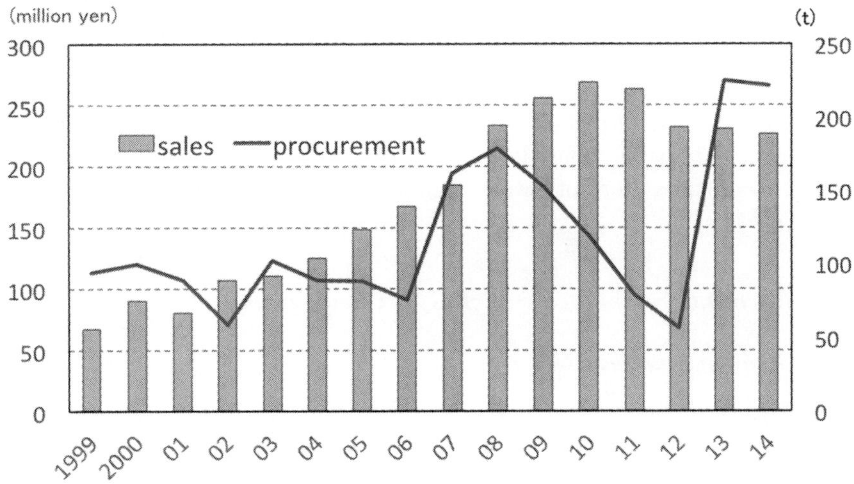

Figure 7.5 Trends in raw procurement volume and product sales for FSP.

Source: Created from FSP documents.

agricultural cooperatives and consumers' cooperatives, which historically worked independently, or even antagonistically. The FSP is a non-profit organization because Japanese laws currently do not allow a form of cooperative that functions both as an agricultural and as a consumers' cooperative. The FSP's success can be partly attributed to the careful crafting of its charter, which specifies the philosophy and rules of the project and assigns specific roles and responsibilities to each cooperative. These rules require growers to comply with specific cultivation standards, food-processing firms to facilitate exchange between growers and consumers, and consumers to support the project through ethical consumption. These efforts helped to sustain the growth of the business under the FSP.

Moreover, the scale of each cooperative and its effects are worth noting. Soy production is coordinated by the Central Union of the Agricultural Cooperatives, which is organized by several agricultural cooperatives in the prefecture. As the Central Union serves as the primary point of contact for the production sector, it can respond quickly to an expansion in consumer demand. On the other hand, while there are a number of consumers' cooperatives in Fukushima Prefecture, it is Co-op Fukushima, the largest consumers' cooperative in the prefecture, that leads the FSP. Co-op Fukushima is headquartered in Fukushima City, has 182,000 members, is capitalized at 6,690 million yen (approximately US$56 million), and sold 20.8 billion yen (US$174 million) worth of products in 2013. Co-op Fukushima invited other consumers' cooperatives within the prefecture to participate in the FSP, and their increased participation resulted in an expansion of subscribers. Having major organizational actors representing both producers and consumers serves as the basis to sustain the project's growth.

On the other hand, it is important for the development of a project not only to expand its scale, but also to deepen the level of integration of producers and consumers through the project. Co-op Fukushima has always been characterized by the strong activism of its members. It has been actively involved in the campaign against genetically modified food, the food-mileage movement aiming to reduce CO_2 emissions by switching from imported to prefecture-grown foods, and the "School of the Field," in which coop members visit soybean farms and engage in farming to deepen their knowledge about food production. One cannot dismiss this organizational culture of active engagement as an underlying factor in the FSP's success.

The Fukushima Soybean Project after the disaster

Effects of the nuclear disaster

The FSP's encouraging prospects were dashed by the Great East Japan Earthquake Disaster. Losses took a number of forms. At the Uchiike Jozo Co.—an FSP member and a manufacturer of *miso* and soy sauce in Fukushima City—for example, one of the soy-sauce processing tanks was partially destroyed by the earthquake, and a large amount of partially made soy sauce was spilled (Figure 7.6). The tsunami submerged and salinized soybean fields along the coast. This damage

Figure 7.6 Uchiike Jozo factory just after the disaster (March 11, 2011).

Source: Provided by Uchiike Jozo Co., Ltd.

was not trivial by any means, but most of the soy-sauce manufacturing facilities were restored in a half year, and the tsunami's damage to the fields was not extensive. As is well known, it was radioactive contamination that resulted in the most extensive and enduring damage.

The radioactive contamination inevitably set back the FSP's activities and its local food system. However, they also demonstrated noticeable resilience. The following sections examine how the FSP dealt with two major challenges after the nuclear accident: raw-material shortages and consumer anxieties.

Response to raw-material shortages

Before the disaster, the FSP used to draw soybeans from three regions of the prefecture: the Soma area, in the north of the eastern coast (Hama-Dori); Fukushima City area, in the central basin (Naka-Dori); and the Aizu area, located in the mountainous western region. The nuclear accident had a significant impact on these source regions, especially because soybeans relatively easily absorb radioactive materials through the root systems. Radioactive cesium of over 50 Bq/kg, the safety limit set by the government, was detected in some of the harvested soybeans in the prefecture in 2011. Because of this, sales of soybeans harvested in the Souma and Fukushima areas were banned, and the Aizu area, about 100 km away from the nuclear power plant, became the only source of local soybeans for the FSP.

Figure 7.5 shows the trends in soybeans procured (as raw materials) in volume. For two years between 2011 and 2012, the volume of raw-soybean procurement dropped sharply due to the effects of radioactive contamination. Because dry soybeans keep well, processing firms were able to use soybeans from previous years without running out of the raw material immediately. However, severe raw-material shortages were in sight after 2013. The project sought to solve the procurement problem by contracting with agricultural cooperatives in new areas.

In the Shirakawa area, in the south-central region of the prefecture, some irrigation channels were damaged by the earthquake, and farmers could not cultivate rice (there was no shipment restriction on rice in the Shirakawa area). The agricultural cooperatives in the Shirakawa area considered cultivating soybeans to replace rice, but starting soybean cultivation anew would require a large amount of capital to purchase specialized machinery and equipment. Those farmers could not afford such expenses, even with some government subsidies and loans from agricultural cooperatives and producers.

Help came from the disaster-affected agricultural cooperatives located in the coastal regions near the Fukushima Daiichi Nuclear Power Plant. Because the farmers in these regions could not resume soybean cultivation any time soon due to severe radioactive contamination, they offered Shirakawa's agricultural cooperatives a lease of their soybean machinery and equipment. Thanks to this offer, Shirakawa farmers could start soybean production with a smaller initial investment, and the FSP was able to procure soybeans from them. Combined with soybeans from Aizu, this collaborative effort enabled the FSP to avoid a major raw-material shortage.

Response to consumer anxieties

Demand for food produced in Fukushima Prefecture dropped drastically after the nuclear accident not only due to the actual radioactive contamination of food, but also, perhaps more importantly, due to consumer anxieties resulting in stigma being attached to Fukushima products. However, sales of FSP products did not decline as drastically as those of many other foods from Fukushima (Figure 7.5). Recent trends in the number of FSP subscribers (Figure 7.7) indicate little sign of decreased membership. Both figures indicate sustained support for the FSP, and I attribute this to the following three factors.

First, the FSP established a rigorous radioactivity-inspection system to provide its subscribers with safe food and a sense of reassurance. The primary objective of the FSP's inspection system was to prevent contaminated foods from entering into the market. The national and municipal governments had already stipulated mandatory testing of food, but this requirement was established primarily for agricultural produce. Also, because the "official" testing of agricultural produce is based on samples, it was difficult to provide consumers with complete reassurance. To provide such reassurance, radiation testing at many stages of production, including processing, was deemed essential. The FSP-member food-processing firms and warehouses decided to implement radiation testing of their products at three different stages: at the time of harvest (mandatory test), at the warehouse (voluntary test), and at the processing facilities (voluntary test) (Figure 7.8). Only those soybeans and soy products that passed the three-fold testing would be shipped to retail stores and subscribers. Moreover, the FSP widely publicized its inspection system through a number of workshops and brochures. These measures taken by the FSP helped to gain the trust of subscribers and other participating members.

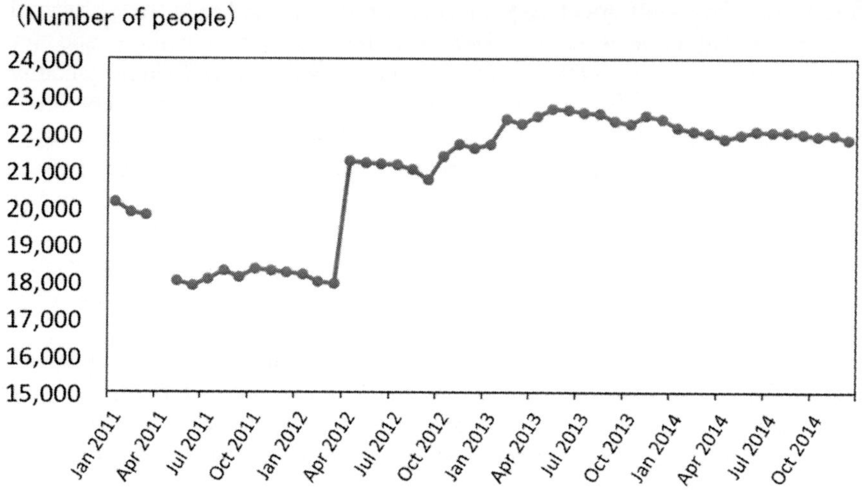

Figure 7.7 Number of members registered with the FSP.

Source: Created from FSP documents.

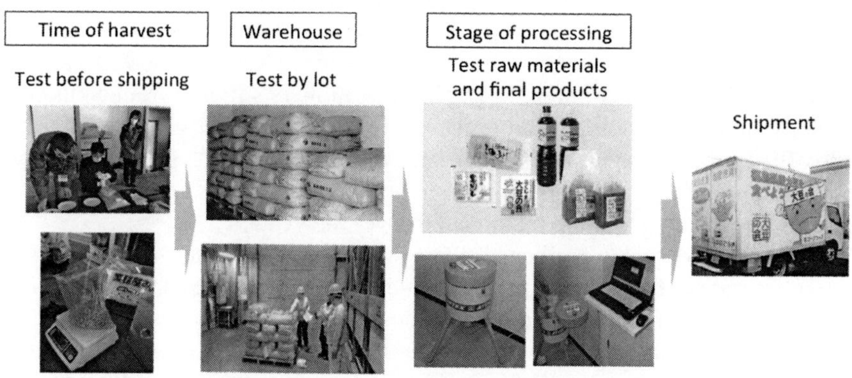

Figure 7.8 Inspection systems at the FSP.

Source: Created from documents by and interviews with the FSP.

Second, some members of consumer cooperatives that participate in the FSP have been voluntarily involved in various radiation-monitoring activities. The subjects of these monitoring activities ranged widely, including soil on farms, vegetables and fruits, diets, and human bodies. First, to measure the level of radiation contamination of farm soils, rather than relying on sparse official monitoring posts that measure only air doses (Chapter 6), the "Soil Screening Project" was set up by agricultural and consumers' cooperatives to measure the concentration level at each single patch of farmland in Fukushima City and other two cities, and to create extremely detailed maps of radioactive contamination. Volunteers for this project came from consumers' cooperatives in and outside of Fukushima. Second, food-radiation monitors were installed at consumers' cooperative offices and their stores in the prefecture, and the members could bring their own food for radiation measurement. These radiation monitors were donated by consumers' cooperatives outside Fukushima. Next, activities to measure radioactive concentration of the contents of a diet as a whole, and to measure bodies' internal exposure (whole-body counter), have been conducted for coop members to obtain more rigorous and accurate information by themselves. In short, the involvement of cooperatives and their members in various radiation-monitoring activities helped them to become cautious but well-informed consumers, ultimately providing them with a sense of confidence in the FSP and in other coop-supplied foods.

Third, beyond these activities that took place after the disaster, we must consider the trust that was cultivated among the FSP members before the disaster as a critical factor providing them with a sense of reassurance. Since its launch in 1998, the FSP has invested a considerable amount of effort into the exchange of ideas, experiences, and emotion among farmers, food-processing firms, and consumers, rather than simply facilitating monetary transactions among them. As consumers, many members of consumers' coops took the trouble to visit farms and factories to learn about producers' challenges and to empathize with their enthusiasm. Producers also made efforts to improve the quality of their products to meet the high expectations of the FSP subscribers. The "School of the Field" allowed members of consumers' coops to experience sowing, harvesting, and making *miso* with their children. In 2007, when farmers suffered from an unusually chilly summer, the coop members supported farmers by actively buying *natto* made of irregularly sized, bruised soybeans with the label of "*ouen* (supportive) *natto*." All these financial, physical, and emotional investments among participating members fostered trusting relationships over the years.

Despite its remarkable resilience, the FSP has not overcome its difficulties. First, it has not yet been able to resume some of the member-exchange activities, such as the "School of the Field," out of consideration for children. Four years after the nuclear accident (at the time of writing), it remains uncertain whether the time has come to resume these activities to bridge consumers and producers, which is seen as an essential part of the FSP. Second, the future direction of the project remains a matter of concern. Although the project's membership did not decline as a result of the disaster, it has not increased substantially either. Annual sales seem to have plateaued at around 200 million yen (US$1,666,000). Both an expansion in volume and a deepening in quality are critical. For an expansion in volume, growth in subscriber numbers would be essential. Currently, the number of subscribers who continuously purchase the project's soy foods is 22,000, which only accounts for 12 per cent of the 182,000 members of Co-op Fukushima as a whole, and there is still room for growth. Meanwhile, it is important for subscribers not merely to participate in the project as one-sided consumers, but to foster a sense that as members they are involved in the solidarity of food and agriculture. For that purpose, plans for exchange and learning programs such as the aforementioned "School of the Field" are required. It might also be a good idea to consider broadening the variety of products offered. A diversity of agriculture, including rice, fruits, vegetables, dairy, and livestock, has developed in Fukushima. The possibilities for local production and consumption are boundless. In fact, *konnyaku*, or konjac, foods were added as a new line a few years ago. Development to foster local production and consumption of a number of the project's items can be anticipated, without being bound by the name of the Soybean Project.

Conclusions

Fukushima has arguably passed the stage of immediate recovery from the disaster, and is now at the point at which strategies for long-term development need to be explicitly charted. As a leading agricultural region of the country, Fukushima's future development cannot be discussed without considering its role in food production. Rather than simply examining the agricultural sector alone, nevertheless, this chapter has focused on food systems, interconnected actors, and processes around particular food commodities. In particular, I have sought to understand and demonstrate how local food systems may remain resilient to such a significant cause of distress as the nuclear disaster. The case study of the FSP shows that the local soybean food system creatively and collaboratively responded to challenges that arose from the radioactive contamination of the food and the environment. Particularly important in the success and resilience of the FSP are the viable business model of support though purchase, the orchestrated synergy between consumer cooperatives and agricultural cooperatives, and the organizational culture of cooperation toward shared goals cultivated among the members since the project's inception in 1998.

The FSP's experiences may provide a model to be replicated in other sectors and commodities. In Fukushima, as in most other parts of Japan, there are many other cooperatives besides agricultural and consumer ones, including fishery and forestry cooperatives. These cooperatives, together with manufacturers, retailers, restaurants—even hospitals and schools—can develop collaborative networks around shared goals to contribute to regional development (Figure 7.1). The main challenge is to develop a business model of win–win relationships, and lessons can be learned internationally from such efforts as fair trade of coffee and bananas (e.g. Tsujimura 2013).

Local food systems, which have drawn attention in various parts of the world, came into existence primarily out of concerns regarding the erosion of local civil society by the forces of globalization—concerns that the architects of the FSP shared. The Great East Japan Earthquake Disaster, and the Fukushima nuclear disaster, turned out to be unexpected and punishing tests of the local food system's resilience to an intense exogenous stress. The future of the FSP remains uncertain and is full of challenges, but without doubt its experience and efforts continue to be critical sources of scholarly inquiry and of inspiration for the rebuilding of Fukushima as a whole.

Acknowledgments

I would like to thank members of the Fukushima Soybean Project for their cooperation during my research. It is my sincere desire to devote myself to continuing my research in the hope of contributing to the recovery of communities and to the revitalization of the food and agriculture industries in Fukushima Prefecture. This study was partly supported by JSPS KAKENHI Grant Number 26892005 (Representative: Takashi Norito).

Notes

1 The concept of food systems has been inspired and informed by such related concepts as commodity systems advocated by Friedland (1984), global commodity chains (GCCs) (Gereffi and Korzeniewicz 1994), and global value chains (Gereffi *et al.* 2001). The GCC framework, in particular—which has been developed on the basis of the world-systems theory of Wallerstein—informs a number of food-system studies, such as Tsujimura (2004).

2 The Act on Interim Measures concerning Subsidies for Soybean and Rapeseed, implemented in 1961, in effect provided subsidies only for those soybeans that were distributed through the national commodity market, and not for those that were sold directly to processing firms. This act was amended substantially in 2007 to allow for more diverse and local distribution channels.

References

Chung, Young-il. 2009. "Current Food Policy Issues in South Korea." *Journal of Food System Research* 16(2): 93–100. [In Japanese]

Fold, Niels, and Bill Pritchard, eds. 2005. *Cross-Continental Agro-Food Chains: Structures, Actors and Dynamics in the Global Food System*. London: Routledge.

Friedland, William H. 1984. "Commodity Systems Analysis: An Approach to the Sociology of Agriculture." In *Research in Rural Sociology and Development: A Research Annual. Focus on agriculture. Vol. 1*, edited by Harry K. Schwarzweller, pp. 221–235. Greenwich, CT: JAI Press.

Gereffi, Gary, and Miguel Korzeniewicz, eds. 1994. *Commodity Chains and Global Capitalism*. Westport: Greenwood Press.

Gereffi, Gary, John Humphery, Raphael Kaplinsky, and Timothy J. Sturgeon. 2001. "Globalisation, Value Chains and Development (Introduction to the Volume's Feature—The Value of Value Chains: Spreading the Gains from Globalization)." *IDS Bulletin* 32(3): 1–8.

Henderson, Elizabeth, and Robyn Van En. 2007. *Sharing the Harvest, Revised and Expanded: A Citizen's Guide to Community Supported Agriculture*. White River Junction, VT: Chelsea Green Publishing.

Hwang, Su-chul. 2008. "The Dynamics of Food System and related Policy in Korea." *Journal of Food System Research* 15(2): 53–58. [In Japanese]

Ito, Koretoshi. 2012. *Regional Revitalization Through Promotion of Local Production for Local Consumption*. Tokyo: Nippon Hyoron Sha. [In Japanese]

Kimura, Aya Hirata, and Mima Nishiyama. 2008. "The *Chisan-Chisho* Movement: Japanese Local Food Movement and its Challenges." *Agriculture and Human Values* 25(1): 49–64.

Koyama, Ryota. 2013. "The Influence and Damage Caused by the Nuclear Disaster on Fukushima's Agriculture." *Journal of Commerce, Economics and Economic History (The Shogaku Ronshu)* 81(4): 11–21.

Lyson, Thomas A. 2004. *Civic Agriculture: Reconnecting Farm, Food, and Community*. Lebanon, NH: Tufts University Press.

Markusen, Ann. 2007. "A Consumption Base Theory of Development: An Application to the Rural Cultural Economy." *Agricultural and Resource Economics Review* 36(1): 9–23.

Morris, Carol, and Henry Buller. 2003. "The Local Food Sector: A Preliminary Assessment of its form and impact in Gloucestershire." *British Food Journal* 105(8): 559–566.

Niiyama, Yoko. 2001. *Gyuniku no Fu-do Shisutemu* [Beef Food System]. Tokyo: Nihon Keizai Hyouronsha Ltd. [In Japanese]

Nishiyama, Mima, and Aya Hirata Kimura. 2005. "Alternative Agro-food Movement in Contemporary Japan." *The Technical Bulletin of Faculty of Horticulture* 59: 85–96.

North, Douglass C. 1955. "Location Theory and Regional Economic Growth." *Journal of Political Economy* 63(3): 243–258.

Okada, Tomohiro. 2005. *Chiikizukuri no Keizaigaku Nyumon: Chiikinai Saitoshiryoku Ron* [Introduction to the Economics of Region-Making: Theory of Intra-Regional Reinvestment]. Tokyo: Jichitai Kenkyu Sha. [In Japanese]

Porter, Michael E. 1998. "Clusters and the New Economics of Competition." *Harvard Business Review* 76(6): 77–90.

Renting, Henk, Terry K. Marsden, and Jo Banks. 2003. "Understanding Alternative Food Networks: Exploring the Role of Short Food Supply Chains in Rural Development." *Environment and Planning A* 35(3): 393–411.

Takahashi, Masao, and Osamu Saito, eds. 2002. *Fu-do Shisutemu Gaku no Riron to Taikei* [Theory and System of Food System Studies]. Tokyo: Association of Agriculture and Forestry Statistics. [In Japanese]

Tansey, Geoff, and Anthony Worsley. 1995. *The Food System—A Guide*. London: Routledge.

Tsujimura, Hideyuki. 2004. *Coffee and North–South Problem*. Tokyo: Nihon Keizai Hyouronsha Ltd. [In Japanese]

Tsujimura, Hideyuki. 2013. *Nogyo wo Kai Sasaeru Shikumi—Fea Toreido to Sansho Teikei* [Mechanism to Support Agriculture through Buying—Fair Trade and Teikei between Producers and Consumers]. Tokyo: Ohta Publishing Co. [In Japanese]

Whatmore, Sarah. 1995. "From Farming to Agribusiness: The global Agro-food Networks." In *Geographies of Global Change: Remapping the World in the Late Twentieth Century*, edited by R. J. Johnston, Peter J. Taylor, and Michael Watts, pp. 57–67. Oxford: Blackwell.

Yamamoto, Daisaku. 2011. "Regional Resilience: Prospects for Regional Development Research." *Geography Compass* 5(10): 723–736.

8 Impacts of the disaster and future tasks for the recovery of small and medium-sized manufacturing firms in Fukushima

Toshio Hatsuzawa

Introduction

The Great East Japan Earthquake Disaster and the subsequent nuclear disaster severely impacted the commerce and industry of Fukushima Prefecture. In addition to the damage to factories caused by the earthquake and tsunami, supply chains were disrupted and many businesses were forced to suspend operations. While the restart of business operations subsequently progressed along with recovery efforts, the pace of reconstruction has varied considerably depending on the nature of the damage caused. In particular, as a result of the accident at the TEPCO Fukushima Daiichi Nuclear Power Plant (NPP), even five years after the start of the disaster, many residents and businesses are still unable to return to the areas around the plant due to high levels of radioactive contamination. In areas to which they are now able to return due to lowered levels of radiation, economic recovery is an urgent matter.

Despite the image of Fukushima as primarily an agricultural prefecture, I emphasize the importance of the recovery of industry and commerce. In Fukushima Prefecture as a whole, more than 92 per cent of workers are in either the secondary or tertiary sectors. In Minami-Soma City, the focus of this chapter and just north of Futaba District, which has been most severely affected by the nuclear accident, 8.2 per cent of the workforce is employed in primary industries, 33.4 per cent in secondary industries, and 58.4 per cent in tertiary industries. Thus, those employed in secondary and tertiary industries comprise more than 90 per cent of the total workforce. In addition, looking at figures for values of production, the agricultural sector totaled 10 billion yen in 2006, while manufacturing output totaled 88 billion yen and retail sales reached 122 billion yen in 2010, clearly indicating that rebuilding local industries is a prerequisite for the region's recovery.

Recovery in the disaster-affected areas requires accurate assessment of the current state of affairs in order to develop effective recovery plans. However, amid the disarray of the affected areas, there has been a lack of even basic data and analyses. Indeed, in the Hama-Dori region of Fukushima Prefecture, which includes Minami-Soma City and Futaba District, even the current state of affairs remains largely unknown. Published reports on the area are limited to: an early report on the conditions of the affected areas immediately after the disaster by Sueyoshi (2011); a field survey of the damage to the manufacturing sector of the Haramachi ward of Minami-Soma and recovery efforts by Hatsuzawa (2012); a study of construction-sector trends in the Haramachi ward following the earthquake disaster by Hatsuzawa (2013); a discussion of the reconstruction of the commercial sector focusing on temporary retail sites conducted by Tsuchiya and Isurugi (2014); and an analysis of tourism and recovery trends in Iwaki City by Muranaka and Tanibata (2012). Although the initial impacts of the earthquake, tsunami, and nuclear accident on businesses have been reported in some journalistic media and scholarly publications in English (e.g. Herod 2011), there have been few English-language publications tracing the (slow) process of industrial recovery in this region. The above-mentioned perception of the marginal status of manufacturing and commerce in this area and the fact that a wide area of the Hama-Dori region remains under evacuation designations may explain the lack of attention by a broader range of scholars.

The following chapter uses a case study of the Haramachi ward of Minami-Soma City, an area located 20–30 km from Fukushima Daiichi NPP, to examine the current state and future tasks of local industries in the nuclear disaster-afflicted areas. Minami-Soma City is an important case study for considering forms of disaster recovery from the nuclear disaster, especially because it is located just outside the areas of above-legal-limit radioactive contamination (hence no formal economic activities are permitted) near the nuclear plant. In many ways, the city is at the forefront of post-nuclear disaster recovery efforts.

The Haramachi ward is located near the geographical center of the Hama-Dori region, approximately 20–30 km north of the TEPCO Fukushima Daiichi Nuclear Power Station. The ward is approximately 200 km^2 in size. The district is situated between the Pacific Coast and the Abukuma Highlands, with a belt of rice paddies in the east and forested mountains in the west. In 2006 the formerly independent Haramachi City was merged with Kashima Town and Odaka Town to form the new municipality of Minami-Soma City. As a result, Haramachi became a ward within Minami-Soma City. The ward is located in the center of the newly formed city, and manufacturing as well as various urban functions are congregated in the ward. The population of the ward just prior to the disaster was 47,000. Since only areas within 20 km of Fukushima Daiichi were designated evacuation zones following the nuclear accident, the Haramachi ward was not an evacuated zone. However, for six months following the disaster, the area was designated an "emergency evacuation preparation area," which resulted in various restrictions, including prohibitions on reopening schools and hospitals. Accordingly, there was a large outflow of population from Haramachi. The pre-disaster population of Minami-Soma in March 2011 was 72,000. This figure remained at 47,000 as of October 2015. Such population loss obviously entails a loss of customers for the retail and service sectors of the city. For the manufacturing sector of the city, population loss results in a decreased workforce and a lack of workers. Moreover, since a number of adjacent towns have been included within the evacuation-designated areas, Minami-Soma's regional sphere of commerce and transaction relations has been reduced and has also been curtailed as a result of stigma (*fuhyo higai*, or "reputational damage"), thus dealing a major blow to the local economy. In great contrast to previous disasters, the nuclear disaster has resulted in an extensive and diverse range of damage to the areas.

The data utilized in this study come from a variety of secondary sources, including local newspapers and published scholarly research in Japanese, four questionnaire surveys conducted in collaboration with the Fukushima Future Center for Regional Revitalization (FURE) at Fukushima University and the Haramachi Chamber of Commerce and Industry, and interviews that I conducted with local businesses. The first survey was conducted in November through December of 2011. The target for the survey was the member businesses of the Haramachi Chamber of Commerce and Industry and responses were received from 168 establishments in the manufacturing sector, which is the focus of this chapter. Subsequently, the survey has been conducted annually (the number of valid responses for the manufacturing sector was 75 in 2012, 84 in 2013, and 102 in 2014).[1] The results of the annual surveys have been published as the *Report of the Survey of Members of the Haramachi Chamber of Commerce and Industry* (Haramachi Chamber of Commerce and Industry and Fukushima Future Center for Regional Revitalization at Fukushima University 2012; 2013; 2014; 2015). These surveys are, to my knowledge, the only such studies that have been carried out on an annual basis in the affected areas following the Great East Japan Earthquake Disaster.

This study reveals that despite their various individual and collective efforts, secondary and lower-tier subcontractors suffered considerably by losing their contractual relations after the disaster. It also identifies that the nuclear accident and subsequent evacuation policies put in place by the government have had more severe effects on local business than the physical damage caused by the earthquake and tsunami. Nevertheless, recovery has progressed since around 2014, even though it has accompanied a growing "recovery gap" among local firms. Finally, the labor shortage which was a major problem in the immediate aftermath of the disaster has been gradually resolved, but local firms are now facing the challenge of securing a high-quality workforce.

Characteristics of the manufacturing sector in Minami-Soma City

Since the manufacturing sector of the Haramachi ward has a rather unique structure, this section will first analyze and explain the pre-disaster structure of manufacturing, in order to frame a later examination of how the disaster radically altered this structure. According to the Fukushima Prefecture Industry Survey, in 2010 there were 202 manufacturing-related businesses in Minami-Soma, with 5,471 employees. Manufacturing-product sales amounted to approximately 89 billion yen, or 1.8 per cent of all manufacturing sales in Fukushima Prefecture. The largest category of sales was electronics, at 16.7 billion yen, followed by electric machinery, at 9.1 billion yen. Thus electronics and electric machinery are the mainstays of the manufacturing sector in this region. However, after the earthquake, manufacturing in the ward rapidly declined, and figures for 2013 show that the number of manufacturing-related businesses had fallen to 150 and the total number of employees to 3,952. Sales of manufacturing products had fallen to approximately 66.6 billion yen (Figure 8.1). The city's contribution to total manufacturing-product sales for Fukushima Prefecture fell to 1.4 per cent, with paper and pulp having the largest sales figure (13.2 billion yen), followed by electronics (11.7 billion yen). Product sales of electric machinery were limited to 3.9 billion yen, thus indicating a drastic transformation in the relative importance of various industries before and after the disaster. The reason for this change is that while some companies have been using reconstruction funds from the government to make large-scale investments, there are also some enterprises that were trading partners with the nuclear power station whose business abruptly stopped. The economic structure of the region has thus been greatly transformed.

It is critical to note, nevertheless, that the decline of manufacturing has been a long-term trend in the region, and the impact of the disaster must be understood in this context. During the 1970s and early 1980s, many Japanese manufacturing firms shifted their production, often in the form of branch plants, to more peripheral parts of the country, such as Tohoku and Kyushu, to access cheaper labor and land. Fukushima Prefecture was relatively close to the Tokyo metropolitan area and attracted many auto, machinery, and electronics plants. Sales of manufacturing products produced in Minami-Soma peaked at 210.2 billion yen in 1991. At that time there were 345 manufacturing-related businesses employing 10,841 individuals, and manufacturing sales accounted for 4.0 per cent of the value of all manufactured goods produced in Fukushima Prefecture. With the rising value of the yen after the Plaza Accord in 1985, and the end of the bubble of

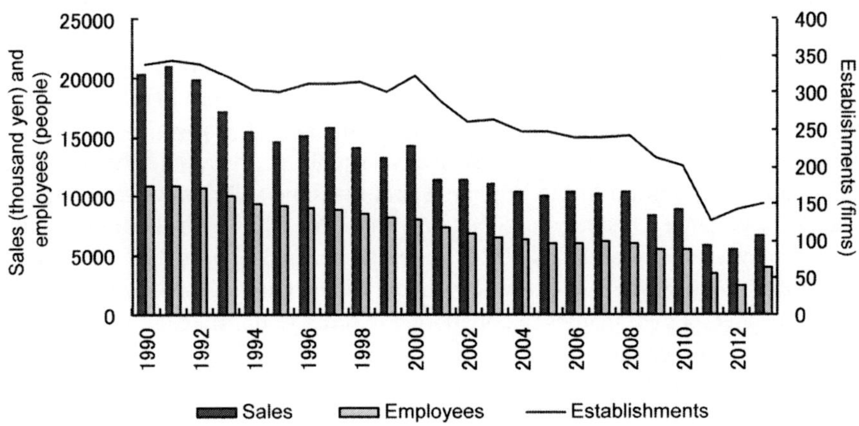

Figure 8.1 Changes in sales, employment, firms in the manufacturing industry in Minami-Soma City.

Source: Based on Manufacturing Industry Survey, Fukushima Prefecture.

economic boom in 1991, large industrial firms were beginning to shift production from domestic peripheries to overseas for further cost advantages. Reflecting this trend, two of the largest manufacturers of electric machinery in Japan had closed their plants in Haramachi by 1993. If we compare 1991 and 2010, just before the disaster, we see a 59 per cent reduction in the number of manufacturing-related businesses, a 50 per cent reduction in employees, and a 42 per cent reduction in product sales. As a result, product sales per business establishment in 2010 were at 71 per cent of 1991 levels.

The structure of the manufacturing sector in the Haramachi ward was composed of a number of subcontractors who produced parts for larger parent companies. When the two large manufacturers of electric machinery withdrew from Haramachi, there was a significant cascade effect on their subcontractors. Many were forced out of business, while others that remained in operation were forced to adapt to the new circumstances. One of the adaptations observed was an increase in what Takahashi (2002) calls "horizontally expanding" (*yokotenkei*) companies, a local term referring to subcontractors that supply their products to multiple parent companies. In the past, only a few subcontractors with extraordinarily high technical capacities were of the horizontally expanding type, while most others dealt with a single parent company. However, the withdrawal of large firms, including those two manufacturers of electric machinery, made it difficult for subcontractors in Haramachi to secure sufficient orders from only one parent company. As a result, a number of subcontractors began to develop transactional relationships with multiple parent companies.

This horizontal expansion of subcontractors can be seen as strengthening their independence *vis-à-vis* parent companies, which also meant that it became more difficult for subcontractors to secure steady orders from their parent companies (as there was less sense of obligation to keep their businesses afloat on both sides). In order to maintain their transactional relations with and bargaining power over parent companies, subcontractors were put under great pressure to improve the quality of their products and cost competitiveness through continuous technological enhancement. It is, however, rather challenging for small to medium-sized enterprises to independently increase their technical and innovative capacities. One way in which local manufacturers have sought to overcome this obstacle is through the formation of the Minami-Soma Machinery Industry Development Council, a voluntary business association formed in 2006 with the aim of supporting member businesses. As of 2015 the council had 23 member companies, including 15 enterprises based in Haramachi ward. One of the council's main aims is for member companies to collaborate in order to improve their production technologies. The council has held various research meetings, not only hosting external lecturers and technical guidance, but also having employees of member companies openly share their knowledge with other member companies. While such knowledge and expertise is usually considered a trade secret to be heavily guarded, the council made the decision to share the information in order to improve the technical capacity of all members.

It is important to note that the council's establishment was not meant simply to enhance member companies' technological capacities; rather, there was another, more practical goal. Following the collapse of the bubble economy, transactional relations became less stable and the duration of these relations became progressively shorter. For example, transactions with one-month contract periods and no guarantee of continuation are not uncommon. The result is that even if large orders are received, it is difficult to increase equipment and staff so as to fulfill such orders. Accordingly, subcontractor companies often had to turn down orders that exceeded their production capacity, even if other conditions, such as profit margins per transaction, were favorable. However, if large orders are shared by subcontractors in the area then it becomes possible to fulfill them. The council was in fact formed under the assumption that large orders that could not be met by one company alone could be fulfilled if production was spread across producers in the area; thus they aimed to secure orders by sharing work. It was for this reason that the council was working to diffuse technical skill and capacity by promoting the open sharing of knowledge among its member companies, so that the member companies would be equally competent to meet the specifications required by their parent companies.

This collaborative strategy played an important role in raising the productive and technological capacities of local industry in the Minami-Soma area during the post-bubble economic boom period, even though the absolute size of the manufacturing industry was on the decline. These efforts were still underway when the Great East Japan Earthquake dealt a major blow to the manufacturing firms in this area.

Damage to manufacturing and recovery of operations following the disaster

The total number of deaths in Minami-Soma City was 1,114 (as of October 2015), including 478 so-called "disaster-related" deaths. This death toll represents approximately 33 per cent of the total 3,798 deaths that resulted from the earthquake disaster in all of Fukushima Prefecture, making Minami-Soma City the most heavily impacted municipality in the prefecture. While the severity of the impact on Minami-Soma City is due in large part to the fact that coastal areas of the city were devastated by the massive tsunami, it is also imperative to emphasize the large number of "disaster-related" deaths. As a result of the prolonged evacuation that followed the nuclear disaster, large numbers of people experienced deteriorations in their health and eventually lost their lives. The impacts of the disaster remain ongoing and are growing even today due to the continuing evacuation of a number of areas.

On March 12, 2011 an evacuation order was issued for areas within a 20 km radius of the nuclear power plant, and the evacuation zone included the Odaka ward of the city (Table 8.1; also see Chapter 5). On March 15 an indoor-shelter order was issued for the areas within 20–30 km of the plant, which included the Haramachi ward. This order remained in effect until April 21. In principle, all business operations, as well as shipping and distribution in the area, were suspended while the indoor-shelter order remained in place. Therefore, although the residents could stay in the area as long as they minimized the time spent outside their houses, there was little economic activity, and hence few jobs to work in or commodities to buy in the area. Accordingly, many residents voluntarily left the area.

Following the earthquake, many enterprises in the Tohoku region were forced to suspend operations as a result of power and water outages, disruption of supply chains, difficulties securing workers, and other challenges caused by the earthquake and tsunami. However, the inland areas that did not experience damage from the tsunami were able to quickly overcome the above obstacles and businesses were able to resume production, in many cases within 1–2 weeks after the disaster (Table 8.2). According to the Tohoku Bureau of Economy, Trade and Industry (2011), among the 123 major manufacturers of the Tohoku region, 46 companies had restarted operations (or in some cases re-opened) by the end of March 2011, 85 companies by the end of April, 96 companies by the end of May, and 105 companies by the end of June.

This early resumption of operations had numerous and diverse effects on subcontractors in the affected areas. Large manufacturers are unable to resume operations if the subcontractors that supply them with parts do not first recover. Following the earthquake, many factories even outside the Tohoku region were forced to suspend operations due to the fact that many manufacturers were dependent on parts manufactured by the small-to-medium enterprises in the Tohoku region. Accordingly, major manufacturers made efforts to spur the recovery of parts manufacturers in the affected areas. However, this support was not provided equally to all parts manufacturers. Sueyoshi (2011, 51) observes that large parent companies have been actively assisting the restart of critical and specialized subcontractors through personal and other direct means, but that the recovery of small

Table 8.1 Changes in the evacuation orders that pertain to Minami-Soma City

March 11, 2011	Evacuation order issued for areas within the 3 km radius from the TEPCO Fukushima Daiichi Nuclear Plant (Minami-Soma City was not included).
March 12	Evacuation order issued for areas within the 20 km radius from the nuclear plant (parts of Minami-Soma were now included).
March 15	Residents in the 20–30 km radius from the nuclear plant were ordered to stay indoors.
April 22	Areas outside of the 20 km radius were reorganized into "deliberate evacuation areas (above 20 mSv/y)" and "evacuation-prepared area (below 20 mSv/y)."
September 30	The "evacuation-prepared area" designation was removed as the condition became more stable.
April 16, 2012	The "deliberate evacuation areas" were reorganized into three groups based on the levels of radioactive contamination.

Table 8.2 Resumption of production by major companies

Date of resumption	*Company name*	*Main products*	*Location*
March 21	Central Motors	Automobiles	Ohira, Miyagi
March 21	Keihin	Auto parts	Kakuda, Miyagi
March 28	Alps Electric	Electronics parts	Six plants in Fukushima and Miyagi
March 28	Tokyo Electron	Semiconductor manufacturing equipment	Oshu, Miyagi
March 28	YKK AP	Construction materials	Ohira, Miyagi
March 28	Keihin	Auto parts	Marumori, MIyagi
March 28	Iris Ohyama	Lighting equipment and others	Kakuda, Miyagi
March 28	Nippon Chemical Industrial	Materials for LCD circuits	Koriyama, Fukushima
March 30	YKK AP	Construction materials	Osaki, Miyagi
April 1	Furukawa Battery	Auto batteries	Iwaki, Fukushima
April 1	Panasonic	Digital camera	Fukushima, Fukushima
April 1	Panasonic	BluRay disk player parts	Natori, Miyagi
April 1	Kanto Auto Works	Automobiles	Kanegasaki, Iwaki
April 5	Asahi Kasei Power Device	Integrated circuits	Ishinomaki, Iwate
April 11	Nippon Paper Group	Paper products	Iwanuma, Iwate
April 18	Hitachi Power Semiconductor Device	Diodes	Minami-Soma, Fukushima

Note: Created by the author based on sources such as Kawakita Shimpo.

and medium-sized subcontractors providing general-purpose goods or processing, without such support, has been slow.

As Sueyoshi (2011) points out, the process of regional industry recovery after the disaster also took advantage of and reinforced this asymmetric power structure among the firms. That is, large manufacturers aiming for a quick recovery in production discontinued transactions with second and lower-tier subcontractors in severely disaster-afflicted areas, and switched their sources of parts products to companies in other regions. This trend was particularly prominent in areas such as Haramachi that were affected by the tsunami and the nuclear accident. According to our interviews in Haramachi, as late as February 2012, 11 months after the disaster, even two of the most advanced manufacturers were able to recover only about 80 per cent of the total number of customer firms with which they had transactional relations prior to the disaster. We must not assume, therefore, that the "recovery" of major manufacturers meant the recovery of local firms or of the production systems that existed prior to the disaster.

The Minami-Soma Machinery Industry Development Council, described earlier, was unable to counter this trend effectively. The system of sharing parts of orders from parent companies across member subcontractors was still a work in progress; furthermore, the drastic reduction in orders was simply too much to bear through the efforts of the council or of individual companies. Under the extraordinary circumstances, each member company was forced to prioritize securing its own orders, and the council was unable to coordinate the sharing of orders.

Case study of the Haramachi ward

Immediate response of the local manufacturing industry

Following the Great East Japan Earthquake Disaster, including the subsequent nuclear accident, nearly all business operations in Haramachi were temporarily suspended. According to a Haramachi Chamber of Commerce and Industry survey, the number of members of this organization that had resumed operations as of the end of June 2011 was limited to 63.0 per cent.[2] Among the 37.0 per cent of businesses that had not yet resumed operations, only 2.6 per cent were unable to resume operations after damage to equipment caused by the tsunami or earthquake; the remaining 34.4 per cent of businesses were unable to restart operations due to the nuclear accident. As a result, not only were operations reduced, but there were also instances in which companies were unable to recover outstanding sales accounts.

The questionnaire survey conducted in fall of 2011 helps reveal the conditions of the local firms immediately after the disaster. First, looking at the damage caused by the earthquake and tsunami, for the 168 business establishments that responded to the survey, 20 had been impacted by a loss of human resources, including the deaths of 29 officers and employees. The rate of lost lives was much higher in the manufacturing sector than in other economic sectors. This is because manufacturing plants were more likely to have been located in lowlands, which were more prone to tsunami damage. Overall property damage, however, was rather limited in Haramachi. Of the 168 establishments, 91, or over half, had incurred property damage, but approximately two-thirds had incurred property damage totaling less than 5 million yen. Only four establishments had incurred property damage totaling over 100 million yen (Figure 8.2).

The fact that property damage was relatively limited allowed many establishments to restart operations quite quickly after the disaster. Of the 146 valid response to the question "When did you resume operation after the earthquake?," 27 businesses stated that they restarted operations in March 2011, while 80 restarted operations in April 2011. The main difficulties resulted from the indoor-shelter order that followed the nuclear accident. As the name clearly indicates, an indoor-shelter order stipulates that residents remain indoors. Accordingly, all

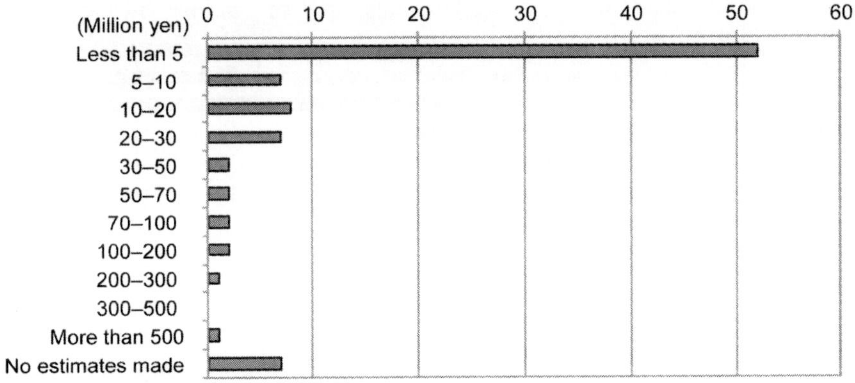

Figure 8.2 Estimated values of damage to properties as a result of the disaster.

Source: Report of the Results of a Survey of Members of the Haramachi Chamber of Commerce and Industry, 2011.

business operations are prohibited as long as an indoor-shelter order remains in place. However, our questionnaire survey revealed that the majority of businesses resumed operations while the indoor-shelter order was still in place. As mentioned above, many of the subcontractors in Haramachi already had precarious relationships with their parent companies, and these subcontractors were pressured to maintain these relationships. This pressure to recover and sustain inter-firm transactional relations forced businesses in Haramachi to resume operations even while the indoor-shelter orders remained in place. Despite these efforts, however, transactional relations were significantly reduced.

This situation was reflected in the response to the question about obstacles for restarting business. "Reduction of orders and customers" tops the list of these obstacles (Figure 8.3). This was followed by "labor shortages" and "evacuation designations resulting from the nuclear disaster." The issues related to labor shortages are critical and will be elaborated further below. Concerns over the evacuation designations imply problems related to stigma. According to our interviews conducted in Haramachi, a number of cases were observed in which transactional relations had been suspended, even when damage from the earthquake and tsunami was relatively minor, due to concerns about the escalating nuclear power plant accident or fears of radioactive contamination, which were solidified and amplified by the evacuation-zone designations. In addition, since shipping companies refused to transport goods through the district, business operators in Haramachi had to travel to Fukushima City and Sendai City to both ship and receive goods, thus forcing a heavy burden upon these operators. In short, it suffices to say at this point that all of these tasks originated in, or were caused by, evacuation and other issues related to the nuclear accident. In Haramachi, the damage from the nuclear accident was far greater than the damage caused by the earthquake or tsunami.

Figure 8.3 Obstacles to the resumption of operation (multiple answers, number of companies).

Source: Report of the Results of a Survey of Members of the Haramachi Chamber of Commerce and Industry, 2011.

Shifts in business indicators since 2011 and a growing "recovery gap"

Our annual survey enables us to monitor and gain important insights into the progress of post-disaster recovery since 2011. The results reveal that the recovery is by no means a linear process. Figure 8.4 presents changes in manufacturing sales in September of each year, with sales in September 2010 representing the 100 figure, and the amounts of capital investment between 2011 and 2013.[3]

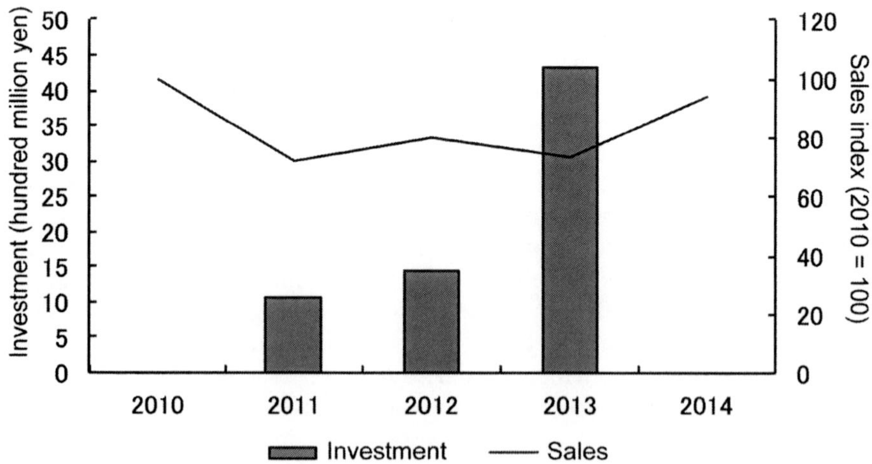

Figure 8.4 Changes in sales and investment among manufacturing firms in Haramachi (sales in 2010 = 100).

Sources: Report of the Results of a Survey of Members of the Haramachi Chamber of Commerce and Industry, 2011–2014 issues.

What deserves special attention here is the rapid expansion of manufacturing sales figures in 2014. As already mentioned, many subcontractors in Haramachi lost their contracts with their parent companies after the disaster, and this is likely the main cause of the stalled recovery of sales through 2013. Does the surge in sales indicate a resumption of the previous transactional relationships? The evidence suggests that that is not the case. Rather, the surge is likely the result of the development of new transaction partners. The survey data indicate that as of September 2014 the value of sales from the transactional relationships that existed prior to the disaster only accounted for 65 per cent of total sales values. Local companies' efforts to develop new markets were starting to pay off after three years of the disaster.

The backdrop to this surge is an expansion in investments from fiscal year 2013 (Figure 8.4). In heavily damaged areas such as Haramachi, it was impossible to pursue sufficient funding and investment for the initial two and a half years after the disaster. Such funding and investment only became possible in 2013. This can be thought to have contributed to the expansion of transactions and sales for the manufacturing sector.

Despite this encouraging finding, the operation of equipment and facilities and the duration of operations have not rebounded to pre-disaster levels. The per centage of establishments that have not recovered or that have reduced utilization rates of equipment and facilities is shown in Table 8.3. Furthermore, approximately one-third of establishments have reduced their sales and operating hours. The average operating time was 2.7 hours in 2012, 2.4 hours in 2013, and in 2.6 hours in 2014. What needs to be emphasized here is that the per centage of businesses that have not returned to pre-disaster use of equipment and facilities changed very little between 2012 and 2014. Facility utilization rates have been reduced by more than 40 per cent for nearly 60 per cent of businesses. For one-third of businesses, operation times have been reduced by nearly three hours.

The operating status of facilities and equipment has worsened especially when compared to sales. This gap is a reflection of growing disparities between businesses that have and those that have not been able to recover from the disaster. In 2014, 8 per cent of establishments were expanding beyond the pre-disaster level of business, and in comparison with businesses that had contracted, this figure was minute. Those establishments whose recovery started slowly appear to be lagging behind further, and are in urgent need of support.

Table 8.3 Capacity utilization rates by manufacturing establishments in Haramachi, compared to the pre-disaster levels

	2012	*2013*	*2014*
Establishments experiencing decline in capacity utilization	57%	60%	57%
Average rates of decline	42%	41%	41%

Sources: Report of the Results of a Survey of Members of the Haramachi Chamber of Commerce and Industry, 2011–2014 issues.

Labor shortages as a bottleneck for manufacturing recovery

One of the critical reasons that recovery has not progressed is labor shortages (Figure 8.3). The results of the surveys show, over the four years since the disaster, that the per centage of businesses that reported having difficulties with labor shortages dropped from 39 per cent (2011) to 25 per cent (2012), but then rose back to 39 per cent (2013) and increased further to 42 per cent (2014). This is indicative of growing labor demand as reconstruction efforts have begun in earnest.

Let us now look at just how serious these labor shortages have been. Table 8.4 presents data pertaining to shifts in the numbers of workers employed by the businesses surveyed. It shows that the number of temporary employees is at 91 per cent of pre-disaster levels and the number of part-time employees at 72 per cent of pre-disaster levels, thus indicating some shortages when compared to pre-disaster levels. For full-time employees, employment has recovered to 97 per cent of pre-disaster levels. Thus it can be thought that the serious labor shortages experienced immediately after the disaster have already been extinguished: this raises the question of why many companies have cited labor shortages.

I suspect that those employers who answered "labor shortages" were in fact suggesting a shortage of high-quality labor. Indeed, according to our questionnaire survey, the per centages of businesses that noted labor-quality problems were 45 per cent in 2013 and 40 per cent in 2014. In contrast, the per centages of businesses that responded that the quality of labor was greater than expected were only 1 per cent in 2013 and 2 per cent in 2014. Although there is no way for us to know what these rates would have been prior to the disaster, the large number of establishments concerned about the quality of labor is still remarkable in our view. Each company has had its hands full attempting to secure adequate quantities of labor, and this is not a situation in which companies can be selective over quality of labor. One of the factory owners that we interviewed pointed out that one of the most skilled workers, who was in charge of the finishing process of their products, resigned after the disaster, and the quality of the products visibly declined. He has not been able to hire an equally qualified worker since. This sentiment was shared by many company representatives that we interviewed.

Table 8.4 Number of employees, by category, of studied establishments, 2011–2014

	Feb. 2011	*Sep. 2013*	*Sep. 2014*
Full-time	1505	1438	1455
Temporary	267	258	244
Part-time	188	127	136

Source: Report of the Results of a Survey of Members of the Haramachi Chamber of Commerce and Industry and Fukushima Future Center for Regional Revitalization at Fukushima University, 2014.

This decline in the quality of labor is also related to employees' work histories. The results of the questionnaire survey for 2014 show that the greatest number of applicants are inexperienced, mid-career job-seekers (Table 8.5). Conventionally, most hires would be from the ranks of recent graduates or mid-career workers with experience, but hiring these types of individuals has been difficult in Haramachi. Based on interviews, we learned that training workers in Haramachi can take more than ten years. Following the earthquake, due to fears regarding radiation and other effects, existing employees in their thirties with young children were the first to voluntarily evacuate the area. These employees were precisely the individuals, with 10–20 years' experience, that were expected to fulfill a central role in workplaces. Even if efforts are made to compensate for this loss of human capital with new hires, there is nothing to do but fill spots with lower-quality, inexperienced labor. The result is a decline in labor productivity, and the companies have been forced to make up for the decline in productivity by hiring more workers, creating a vicious cycle.

Wages have also displayed unique trends. Table 8.6 depicts average wage levels for September 2014. Here, in order to compare with other industries, the results are shown for all sectors investigated. While construction and civil engineering have increased wages by significant margins, the areas of manufacturing and wholesale have only slightly increased and, in contrast, retailers and other services have declined. Conventionally, when labor shortages occur, wages should increase. The reason that this is not initially seen in this case is that many businesses are expecting a downturn after reconstruction-induced demand fades and are therefore suppressing hires of full-time employees, while aiming to respond to immediate demand by using temporary and part-time employees. As a result, if temporary workers in manufacturing are excluded, wages for temporary and part-time workers for all sectors have greatly increased. This business strategy of suppressing hires of full-time workers and raising the wages of temporary and part-time workers also has the effect of making it difficult to hire highly skilled labor. In order to pursue hires of skilled labor, it is imperative to raise the salaries of full-time workers. Of course, this is more easily said than done.

For businesses where orders have not yet fully recovered, it would not be feasible to increase the number of full-time employees and raise wages. Many of the critical skills required for work cannot be gained through formal education, and must be internalized by workers over a long period of time. Furthermore, many

Table 8.5 Previous work experience of employees who were hired after the disaster, 2014

Newly graduated from schools	*Mid-career, with previous experience*	*Mid-career, without previous experience*	*Rehiring of previous employees*	*Others*
16	27	30	18	4

Source: Report of the Results of a Survey of Members of the Haramachi Chamber of Commerce and Industry and Fukushima Future Center for Regional Revitalization at Fukushima University, 2014.

Table 8.6 Wage differentials before and after the disaster

	Before the disaster			*After the disaster (Sep. 2014)*		
	Full-time (Monthly, 10,000 yen)	*Temporary (Monthly, 10,000 yen)*	*Part-time (Hourly, yen)*	*Full-time (Monthly, 10,000 yen)*	*Temporary (Monthly, 10,000 yen)*	*Part-time (Hourly, yen)*
Manufacturing	22.2	16.1	758	22.6 (+0.4)	15.2 (–0.9)	851 (+93)
Construction/ civil engineering	25.1	19.2	840	27.8 (+2.7)	24.0 (+4.8)	973 (+133)
Wholesale	19.2	NA	751	19.3 (+0.1)	NA	875 (+124)
Retail	20.5	9	741	20.4 (–0.1)	13.2 (+4.2)	815 (+74)
Other services	22.7	17.3	794	23.3 (+0.6)	20.5 (+3.2)	866 (+72)

Source: Report of the Results of a Survey of Members of the Haramachi Chamber of Commerce and Industry and Fukushima Future Center for Regional Revitalization at Fukushima University, 2014.

of the local firms cannot currently afford to funnel their resources and capital into labor training. Dealing with the lack of quality labor may require creative and potentially collective strategies, as seen in the formation of the Minami-Soma Machinery Industry Development Council almost a decade ago.

Conclusion

This chapter has examined the conditions and trends of industrial recovery and reconstruction following the Great East Japan Earthquake Disaster through a case study of the manufacturing industry in the Haramachi ward of Minami-Soma City. Sales have come close to returning to pre-disaster levels. However, even four years after the disaster, from various perspectives the effects of the disaster are still great and recovery remains insufficient.

This chapter reveals that in order to assess the process of industrial recovery from the disaster, it is essential to understand the conditions of the industrial production systems prior to the disaster. Haramachi's manufacturing sector had the unique characteristic of having many subcontractors reliant on a few large parent companies. Despite individual and collective efforts since the 1990s to diversify their customer firms (i.e. horizontal expansion) and enhance their technological capacities, secondary and lower-tier subcontractors suffered considerably from the loss of their contractual relations after the disaster.

The annual surveys have also revealed that many of the obstacles to restarting business were caused not so much by the earthquake or tsunami damage, but more by the nuclear accident and the government's subsequent evacuation policies. Despite these difficulties immediately after the disaster, our more recent surveys indicate that recovery has progressed since around 2014. Enabled by increased capital investment, local companies have been able to increase the volume of production and make efforts to expand their markets, with new customer firms. Yet, there is also a growing "recovery gap" between companies that are recovering and those that are not. One of the causes of this growing gap can be attributed to the differences in the technological levels of the local firms. The disaster led larger parent companies to seek alternative subcontractors outside of the disaster-afflicted areas. For relatively simple products, these companies decided to maintain the newly formed transactions even after the disaster, making the recovery of some Haramachi-based subcontractors more difficult. Subcontractors with a critical technological edge were able to reestablish business connections with their parent companies. In other words, the disaster accentuated the pre-disaster difference in business performance based on technological levels among firms.

Labor shortages have been identified as one of the reasons for the stalled progress of industrial recovery. However, the number of workers has in fact recovered to near pre-disaster levels. I suggest that the problem of labor shortage is in fact a problem of labor quality. This is in part related to the national government's reconstruction schemes. A large amount of public investment was made for the recovery of the disaster-afflicted areas initially, but, knowing that the investment would last only for a short period, many local companies have been hesitant to hire full-time, permanent workers at high wages. Therefore, I expect that it will take a long time and hard work to improve the quality of the workforce in Haramachi, and believe that some form of public assistance, which may take the form of an industrial-research institute and a new public university, is needed.

Acknowledgments

This work was supported by JSPS KAKENHI Grant Number 25220403 and FURE, Fukushima University.

Notes

1 In the 2011 report the manufacturing and construction sectors are combined under the title "manufacturing." Since they are separated in the reports from 2012 onward (which also explains the smaller response sizes from 2012), the results from 2011 lack continuity with later reports.
2 According to an interview conducted by the author, there are many businesses that have re-opened but where operating rates are low, and the operating rate in October 2011 was estimated to be 40 per cent.
3 The characteristics of the survey mean that caution must be applied when handling the cumulative figures due to changes in the number of businesses responding each year.

References

Haramachi Chamber of Commerce and Industry and Fukushima Future Center for Regional Revitalization at Fukushima University. 2012. *Report of the Results of a Survey of Members of the Haramachi Chamber of Commerce and Industry, 2011*. [In Japanese]

Haramachi Chamber of Commerce and Industry and Fukushima Future Center for Regional Revitalization at Fukushima University. 2013. *Report of the Results of a Survey of Members of the Haramachi Chamber of Commerce and Industry, 2012*. [In Japanese]

Haramachi Chamber of Commerce and Industry and Fukushima Future Center for Regional Revitalization at Fukushima University. 2014. *Report of the Results of a Survey of Members of the Haramachi Chamber of Commerce and Industry, 2013*. [In Japanese]

Haramachi Chamber of Commerce and Industry and Fukushima Future Center for Regional Revitalization at Fukushima University. 2015. *Report of the Results of a Survey of Members of the Haramachi Chamber of Commerce and Industry, 2014*. [In Japanese]

Hatsuzawa, Toshio. 2012. "Some Problems of the Manufacturing Revival From the East Japan Great Earthquake Disaster Stricken Area: A Case Study of Haramachi Area, Minamisoma City, Fukushima Prefecture." *Regional Economic Studies* 24:29–37. [In Japanese]

Hatsuzawa, Toshio. 2013. "Nuclear Disaster and Regional Industries." *Nihon no Kagaku sha* [Scientists in Japan] 48(10): 48–53. [In Japanese]

Herod, Andrew. 2011. "What Does the 2011 Japanese Tsunami Tell us About the Nature of the Global Economy?" *Social & Cultural Geography* 12(8): 829–837.

Muranaka, Akio and Go Tanibata. 2012. "Recovery and Reconstruction of the Tourism Industry After the Great Eastern Japan Earthquake Disaster." *Rekishi Toshi Bosai Ronbun Shu* [Journal of Disaster Mitigation for Historical Cities] 6: 377–384. [In Japanese]

Sueyoshi, Kenji. 2011. "Regional Characteristics of Damages to Industries and Their Recovery." *Chiri* [Geography] 56(10): 48–54. [In Japanese]

Takahashi, Toru. 2002. "Corporate Behavior During Economic Recession and the Restructuring of Local Manufacturing: The Case of the Northern Part of the So-So (Soma-Futaba) Region in Fukushima Prefecture." Unpublished MA Thesis, Fukushima University. [In Japanese]

Tohoku Bureau of Economy, Trade and Industry. 2011. *Sangyo Fukko Akushon Puran Tohoku* [Industrial Reconstruction Action Plans Tohoku]. Accessed June 30, 2016. www.tohoku.meti.go.jp/kikaku/vision/pdf/11fukkou.pdf. [In Japanese]

Tsuchiya, Jun and Shinobu Isurugi. 2014. "Local Commerce in Minami-Soma City After the Great Eastern Japan Earthquake Disaster and the Nuclear Accident." In *Examination of Post-disaster Reconstruction Policies and Proposals for New Industrial Creation*, edited by The Research Project on Regional Industrial Reconstruction, Department of Economics, Tohoku Revitalization University, pp. 148–169. Sendai: Kawakita Shinpo Publishing Center. [In Japanese]

9 Bringing businesses back, bringing residents back

Efforts and challenges to restore commerce in formerly evacuated areas

Akira Takagi and Masayuki Seto

Introduction

The Great East Japan Earthquake Disaster and the subsequent TEPCO Fukushima Daiichi Nuclear Power Plant (NPP) accident forced the comprehensive long-term evacuation of an extensive area centered primarily on the eight towns and villages of Futaba District, along the Pacific coast of Fukushima Prefecture. This long-term, all-resident evacuation has negatively affected the culture, lifeways, and socio-economic functions of each locality. Moreover, in the nuclear disaster-afflicted areas, the dangerous air radiation-dose rates make it highly difficult to advance the reconstruction and recovery of critical infrastructure. Even in cases where return to these evacuated villages has become possible, the progress of recovery and reconstruction has been sluggish.

This disaster also forced a large number of evacuees to relocate from coastal and mountainous rural districts to larger cities and towns. Individuals who evacuated to these more conveniently situated urban communities are bound to experience a drop in services and facilities when returning to their hometowns. This is certainly one reason for the slowed progress of resident return to evacuated areas. Accordingly, in municipalities where return is now possible, it is imperative to take proactive measures to improve the accessibility of medical, commercial, and transportation services to better than pre-disaster standards in order to draw residents back.

It is in this context that the concept of Build Back Better (BBB) may be usefully applied to the recovery and reconstruction processes in Fukushima. The idea of BBB first emerged during the recovery effort following the Indian Ocean tsunami (Clinton 2006), and has since become a key conceptual tool within the post-disaster recovery and reconstruction process (e.g. Khasalamwa 2009; Kennedy *et al.* 2008). BBB is defined as "a way to utilize the reconstruction process to improve a community's physical, social, environmental and economic conditions to create a more resilient community" (Mannakkara and Wilkinson 2014, 319). In other words, the BBB approach focuses on understanding the process of post-disaster recovery and reconstruction from the perspective of victims who wish to return to disaster-afflicted areas. BBB has been extensively utilized in empirical research. For example, Lyons (2009) analyzes the reconstruction of housing in Sri Lanka after the Indian Ocean tsunami from the perspective of BBB. Fan (2013) has examined the cases of Aceh, Indonesia, and Sri Lanka after the Indian Ocean tsunami, the 2008 cyclone that devastated Myanmar, and the 2010 Haiti earthquake to identify the various roles of individuals involved in BBB. Despite its growing popularity in disaster research, the actual practice of BBB has also revealed various challenges (Mannakkara and Wilkinson 2014, 336–337), including: balancing structural improvement with affordability, time constraints, and preferences and traditions of the local community; coordinating the interests of various stakeholders without losing programmatic efficiency; establishing reconstruction programs that take advantage of available skills, resources, and future demands; and achieving speedy recovery and reconstruction results without compromising quality. Accordingly, studies that examine the effectiveness of BBB as a principle of practical guidance for post-disaster recovery and reconstruction must pay close attention to these latent problems and challenges.

The potential value of the BBB concept and practice is being recognized in Japan (e.g. Chapter 2). However, few empirical studies have adopted this perspective to examine areas afflicted by the Great East Japan Earthquake Disaster, and virtually none have focused specifically on areas afflicted by the TEPCO nuclear accident. In this chapter we analyze efforts to practice BBB in areas compulsorily evacuated as a result of the nuclear disaster. By doing so we wish to articulate prospects and limitations of the BBB approach that have not been clearly recognized in the existing BBB literature, which typically focuses on recovery efforts following a natural disaster such as a tsunami or earthquake. The creation of excluded spaces (e.g. restricted zones) due to radioactive contamination strongly shapes the possibilities for BBB following a nuclear disaster.

We focus on Kawauchi Village, a village in the Futaba District of Fukushima Prefecture (also see Chapter 4). Kawauchi Village is situated southwest of Fukushima Daiichi, within 20–30 km of the reactors. Following the accident at Fukushima Daiichi NPP, the entire village was designated an evacuation area and later an emergency evacuation preparation area. It was one of the villages from which the entire population was required to evacuate. However, since the radiation contamination of this village was not as severe as that in nearby municipalities, the village became the first comprehensively evacuated municipality to issue a "Declaration to Return" in February 2012. The early return of residents is one important reason to focus on Kawauchi Village. Another is that the community is currently working to restore the area's social facilities following residents' return after nearly a full year of evacuation. Recording and analyzing the reconstruction process and experience in this village will prove useful for other villages when residents begin to return from evacuation.

Our research methods include analysis of statistical data and local policy documents, as well as interviews with personnel of government organizations and the managers of food-product retailers. Of the eleven food retailers that had been in business prior to the disaster, six had reopened in the village by July 2012. As part of the joint research project with the Fukushima prefectural government, we conducted interviews with these six retailers and with one who had not reopened at the time of our study.

Impacts of the disaster on Kawauchi Village and the post-disaster situation

Although the shaking from the earthquake in Kawauchi Village was quite strong (i.e. 6-upper on the Japanese Meteorological Agency's seismic intensity scale), the damage caused by the earthquake was relatively minor. The village government opened a disaster-response center and initiated attempts to respond to the earthquake. However, although nearly the entire village is located within 30 km of Fukushima Daiichi, the number of people who thought the nuclear accident would seriously impact the village was small. This may, at first, seem somewhat odd or even naïve. One might expect that a village so close to Fukushima Daiichi would be immediately concerned about the impacts of the earthquake and tsunami on the nuclear power plant and the threat of a nuclear accident. However, Kawauchi Village is topographically isolated from Tomioka Town on the coast by mountainous terrain; hence, Fukushima Daiichi has historically been somewhat distant physically and mentally from village life. Accordingly, even after news of the nuclear accident arrived, few people thought that it would have a serious effect on the village.

Contrary to these initial expectations, however, the nuclear disaster worsened with every passing day and did eventually have very large impacts on the village. The day after the earthquake, the Futaba Police Headquarters was evacuated from Tomioka Town, and approximately 8,000 of the town's residents were evacuated to Kawauchi Village. Since the pre-disaster population of Kawauchi Village was only 3,000, such a large inflow of evacuees strained the local government's ability to respond to the situation. Even more problematic, following the hydrogen explosion at Fukushima Daiichi Reactor 4 on March 15, the hourly radiation-dose rate in the village reached 20.5 μGy (≒ μSv), up to 300–500 times the normal level.[1] The village government closely observed the worsening situation and on March 16 made the decision to evacuate the entire village to a convention center, *Big Palette Fukushima*, in Koriyama City,[2] in tandem with residents and officials from Tomioka Town.

In April 2011, secondary evacuation was initiated and evacuees were relocated from the convention center to hotels and inns. In May, evacuees from Kawauchi Village were allowed to make temporary returns to their homes in the evacuated areas. In June, evacuees began to move into emergency temporary housing in Koriyama City as construction was completed. Evacuation conditions were thus gradually improved between March and June 2011. When reconstruction plans were publicly announced by the local government in September 2011, it was also announced that evacuation areas were to be re-designated and evacuation orders lifted at the end of the month for most of the areas within the 30 km radius of Fukushima Daiichi (including Kawauchi Village) that had been "emergency evacuation preparation areas."

Following the lifting of evacuation orders, the village began initiating efforts to return evacuees to the area. In February 2012, Mayor Endo issued a "Declaration to Return," calling on individuals who were able to do so to lead the return amid ongoing and contentious debate regarding the potential effects of radioactive contamination on villager health. At the end of March 2012, village offices were returned to Kawauchi Village from their evacuation site in Koriyama City, and in April, schools and clinics were reopened and a new bus line established. Generally speaking, then, various functions within the town began to be restored. As the advance guard for this process of returning residents to previously evacuated areas, the former "emergency evacuation preparation areas" drew much attention from officials and victims hoping to understand the many issues involved in this process. The all-resident evacuation of Kawauchi Village lasted nearly one year. The declaration to return came rather quickly compared with other towns and villages that had likewise experienced all-resident evacuation. Yet, despite the relatively short period of evacuation, it became clear that residents were not returning apace even after the declaration to return was announced.

As can be seen from the data presented in Table 9.1, based on Basic Resident Register (*Jumin Kihon Daicho*) data,[3] the population of Kawauchi Village today remains largely unchanged from before the earthquake. While a casual glance at this data gives the impression that residents have returned to the village, what it really indicates is that residents of Kawauchi Village have simply not changed their official place of residence due to concerns about whether they will be eligible for compensation for damages resulting from the nuclear accident. What these data also reveal is that the *number of households has increased* since the disaster. Before the disaster, mountainous rural districts such as Kawauchi Village were populated by a large number of multi-generational households. During the course of the evacuation and relocation to temporary housing it was not possible, in many cases, to secure housing large enough to accommodate such households. The Disaster Relief Act sets the standard size for emergency temporary housing at 29.7 m^2. The typical layout is generally two bedrooms with a dining room and kitchen. While evacuees residing in government-subsidized evacuation housing are often living in rental units large enough to accommodate families, there are very few options for accommodating three-generation households. Furthermore, and this seems to be a characteristic specific to a nuclear disaster, concerns for

Table 9.1 Population change in Kawauchi Village

	2010 (Before disaster)	*2012*	*2013*
Population	2,820	2,811	2,805
Number of households	950	1,113	1,132
Average number of people per household	2.97	2.53	2.48

Sources: Census of Japan 2012, Basic Resident Registers 2012 and 2013.

Note: The evacuation in Kawauchi Village lasted from March 2011 through February 2012 (11 months).

children with regard to radioactive contamination have led to the division of families by generation—often a tripartite division between grandparents, fathers and mothers, and children. All of these factors have resulted in increases in the number of (smaller) households.

Approximately 80 per cent of the evacuees from Kawauchi Village were in evacuation in Koriyama, Iwaki, and Tamura cities in July 2012, and approximately 60 per cent of these evacuees were living in Koriyama City (Figure 9.1). Some of these evacuees and households have returned to Kawauchi Village since then. However, accurately identifying the number of individuals and households that have returned to Kawauchi Village is a task fraught with difficulty. The village government is estimating the number optimistically by counting individuals who reside in the village for more than four days a week as "returnees." While in April 2012 returnees numbered around 500, as of October 2012 this figure had increased to 1,100, or slightly more than 33 per cent of the pre-disaster population.[4]

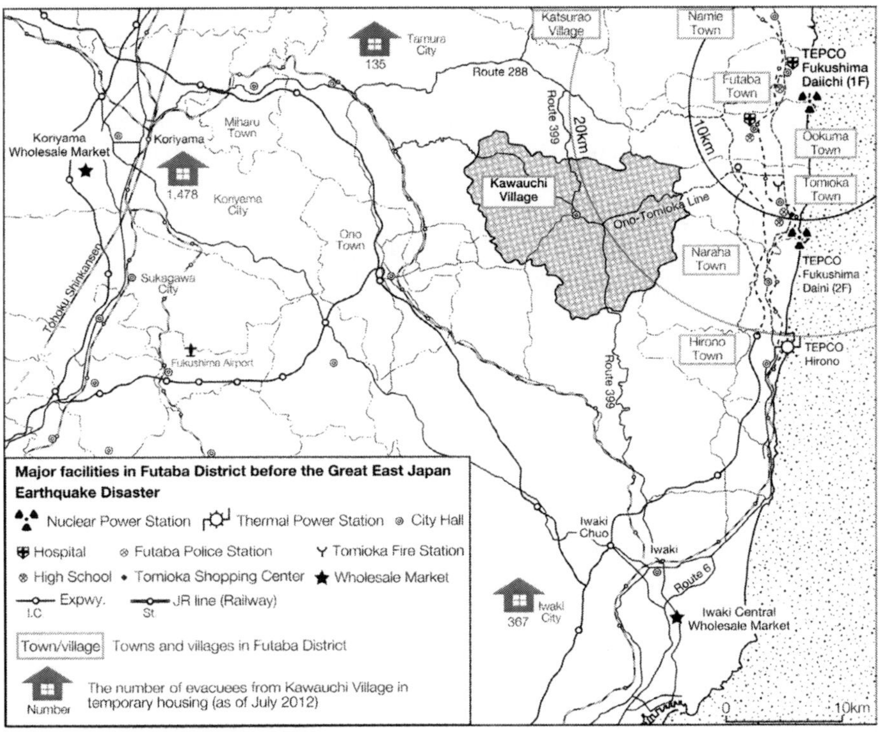

Figure 9.1 Major facilities around Kawauchi Village before the Great East Japan Earthquake Disaster.

Note: Created by the authors.

The questionnaire survey we conducted with the village office revealed why former residents were hesitant to return. These reasons include the village's proximity to Fukushima Daiichi, the ongoing status of the accident, soil contamination, other concerns stemming from the accident, labor issues, school and hospital issues, and also issues regarding shopping opportunities and other daily conveniences.[5] Among the many reasons identified, the issue of shopping opportunities should be given careful attention, both because it is not one that is frequently discussed in the literature as a barrier to return and because this is one of the few factors which local stakeholders can effectively work to resolve.

Interviews with local officials and businesses also revealed why shopping opportunities turned out to be a critical barrier for return. Most villagers evacuated to Koriyama City, a large urban area that is far more convenient and well supplied with urban amenities than Kawauchi Village. Elderly individuals who had depended on friends or relatives to take them shopping or to hospitals while residing in Kawauchi Village were now able to make such trips on their own. Temporary housing sites in Koriyama are situated in convenient neighborhoods with easy access to shopping—conveniences highly appreciated by younger generations. The degree of convenience found at the evacuation sites stands in marked contrast to the many inconveniences of village life.

It has become difficult to obtain fresh food in Kawauchi Village. Although situated in a mountainous rural district, Kawauchi Village is also only 22 km from the coast, and fresh fish and other fresh food were readily available in the village before the disaster, through shops in the village. Following the disaster, fresh fish was no longer available at shops within the village and could not be purchased even at nearby shops outside it. Village residents now need to take much longer trips—perhaps only once a week—to obtain any fresh fish, a fact highly indicative of how drastically the retail facilities of the village have been curtailed.

The village government is attempting to take proactive measures to respond to these issues and bring residents back to the village. The Fourth Comprehensive Plan of March 2013 notes the need to improve the environment of the area and to provide services, including commercial functions, demanded by residents.[6] We can also interpret this move as articulation of the goal of BBB in the specific context of post-disaster Kawauchi Village.

A severed regional commerce network

Unlike most natural disasters, where recovery and reconstruction can start immediately after the event, severe nuclear accidents create spaces of exclusion where recovery is suspended for a long period due to radioactive contamination. The Fukushima nuclear disaster resulted in such spaces: designated evacuation areas around the Daiichi NPP. Although most of Kawauchi Village is now outside of these areas, the effects of their presence are still felt because the village is embedded in a larger regional system, as described below. This has important implications for implementing BBB in the village.

Before the earthquake and nuclear disasters, life in Kawauchi Village was closely connected to the coastal corridor of Futaba County, and particularly to the neighboring community of Tomioka Town (see Figure 9.1). In other words, Kawauchi Village and Tomioka Town were part of the same economic area (or "functional region," to use a classic geographic term). Kawauchi Village was highly dependent on key social facilities in Tomioka Town, including two hospitals; a shopping center; a train station that served as a hub for daily journeys to schools in Futaba Town and Iwaki City; and also employment at TEPCO, its contractors, and affiliated companies in the town. Tomioka Town was essentially the gateway for travel into and out of Kawauchi Village. Residents of Kawauchi Village stopped in Tomioka Town to shop on their homeward journeys, and the neighboring town offered Kawauchi Village residents numerous critical services. In addition, in terms of shipping and distribution connections, Kawauchi Village was positioned as a stopover point between Koriyama City in the Naka-Dori region (i.e. the "central corridor" region of Fukushima Prefecture) and Iwaki City in the Hama-Dori region (i.e. the "coastal corridor" of Fukushima Prefecture). Almost all of the village's shopkeepers traveled via Tomioka Town to the wholesale markets along the coast and in Iwaki City to purchase stock. Additionally, the village's administrative capacities are tied to the wider regional government and administrative networks of Futaba County and, accordingly, many of its police, fire, and waste disposal services are part of a countywide system.

As a result of the nuclear disaster, however, the towns and villages of Futaba District were completely evacuated and the communities along the coast were designated "restricted areas." Accordingly, the regional commerce network in which Kawauchi Village is embedded was severely crippled. The many services and capacities that these towns had provided were brought to a halt and the vital linkages that bound Kawauchi Village to the region were severed. Thus, the process of bringing evacuated residents back to Kawauchi Village has also necessarily been a process of recovering and rebuilding a wider regional system that was temporarily destroyed (Seto and Takagi 2014).

Commercial services and capacities are not exceptions to the above-noted processes of regional socio-economic disruption, and both distribution and shipping into Kawauchi Village have been impacted by the disaster and evacuation. Table 9.2 presents data obtained during interviews with managers of food-product retail shops in Kawauchi Village. From these data we see that stores C and E have suspended the sale of fresh foods and that store G has ceased operations entirely. Moreover, each of the managers interviewed noted that their store has been affected by changes in distributors and difficulties in securing goods. Various disruptions to distribution and shipping were indicated by store managers. For example, in regard to receiving stock, the manager of store A indicated that shipping companies refuse to deliver, the manager of store B noted that some wholesalers no longer come to the area, and the managers of stores F and C noted that bread-delivery services to their stores had been discontinued. The managers of stores B and E noted that former wholesale companies had gone out of business and that there were many financial obstacles to starting business relations with new wholesalers. The number of customers has decreased at each store since the disaster, a fact indicative of the harsh conditions under which businesses in the area must operate. In addition, some store managers expressed concerns that evacuated residents have become accustomed to the ease of shopping at large shopping malls and mega-stores in cities.

At present, the key question being asked of commerce in Kawauchi Village is whether the return of residents must come before the recovery of commercial services and capacities, or whether these services and capacities must be enhanced before residents will return. Without sufficient demand for goods, it is difficult for retailers to enhance their retail services. On the other hand, for residents weighing whether to return to the former evacuated areas, the enhancement of retail services represents a prerequisite to returning. As noted above, before the disaster, residents of Kawauchi Village traveled to a shopping center in Tomioka Town to obtain goods that could not be purchased within the village. Residents must now purchase such goods at large shops in Koriyama and Tamura City, two of the most prominent evacuation destinations.

Prior to the earthquake disaster, Kawauchi Village was situated within a shipping and distribution network that crisscrossed and connected the Hama-Dori and Naka-Dori regions of Fukushima Prefecture. This network was severed by the earthquake disaster, and reopening the numerous truncated dead-ends that appeared in this network is a key task for moving forward. Restoring shipping and distribution connections to a village that has lost two-thirds of its population is no easy task. For example, before the earthquake disaster, one bakery chain baked bread in Koriyama City and then shipped it to Iwaki City and Tomioka Town; the delivery trucks would then stop in Kawauchi Village on their return trip to Koriyama. After the earthquake, however, as a result of Tomioka Town being designated a restricted area, this route was discontinued, as were all direct deliveries to Kawauchi Village. Accordingly, this same bread must now be purchased for higher prices at the wholesale market in Iwaki City. This also results in increased labor costs for securing stock at each store, and lowering these costs is a major issue that must be resolved in the future. Prior to the earthquake disaster it was possible to travel through the coastal corridor to secure stock. However, retailers must now travel to Iwaki City and Koriyama City to procure their goods (Figure 9.1).

Efforts to rebuild commercial services and capacities

With conditions for recovery restricted due to the nature of the nuclear disaster, various stakeholders in Kawauchi Village seek to restore and improve the quality of life of returnees. In terms of commercial function and capacity, we can identify at least three distinct implementations of BBB in the village. First is a joint purchasing program initiated by the Japanese Consumers' Co-operative Union (Co-op).[7] Some Kawauchi residents, while living in temporary housing in Koriyama City, were using a "joint purchasing system" in which a group of Co-op members jointly signed up for regular food deliveries. When they returned to Kawauchi Village, these residents consulted with the Co-op office to consider extending this system to the village. An agreement was reached in October 2012. Initially 33

Table 9.2 Changes in business conditions of retail stores in Kawauchi Village (August, 2012)

Store	*Items sold/stocked*		*Timing of reopening*	*Wholesaler*		*Remarks*
	Before disaster	*After disaster*		*Before disaster*	*After disaster*	
A	Groceries, household goods, liquor, propane gas	Unchanged	Reopened in Aug 2011, but had supplied gas in the immediate aftermath of GEJE.	Food from Tomioka; household goods from Niigata Pref.; propane gas from Minami-Soma and Iwaki	Food from Koriyama	Some transportation companies no longer deliver goods to Kawauchi
B	Groceries, household goods, liquor, fuel, gas station	Unchanged	Reopened in July 2011	Household goods from Koriyama and Iwaki; liquor from Koriyama, Iwaki, and Namie; gasoline from Tokyo	Household goods from Iwaki; liquor from Koriyama and Iwaki; gasoline from Tokyo	Wholesalers no longer come to Kawauchi. It is difficult to contract with new wholesalers
C	General store, perishables	Suspended perishables	Reopened in June 2011; store owner did not evacuate	Perishables from Tomioka	Confectionery from Ono Town	The wholesaler in Tomioka Town has discontinued its business. Bread delivery was suspended
D	Convenience store	Suspended magazines	Reopened in May 2011, but suspended in Aug 2011; reopened as a new franchised store in Nov 2011	Point-of-sale system	Point-of-sale system	The store used to be a retailers' cooperative store. After the disaster, a large national convenience store franchise chain opened the store as a form of post-disaster support

(Continued)

Table 9.2 (Continued)

Store	*Items sold/stocked*		*Timing of reopening*	*Wholesaler*		*Remarks*
	Before disaster	*After disaster*		*Before disaster*	*After disaster*	
E	Fresh fish, fresh meat and catering	Suspended perishables and catering	Reopened in March 2012	Perishables from Tomioka, Namie, and Iwaki	Processed food from Koriyama	Reopened the store in response to numerous requests by the villagers
F	Perishable foods and groceries	Unchanged	Reopened in April 2012	Wholesaler market in Iwaki	Wholesaler market in Iwaki	Delivery of baked items is discontinued
G	Perishable foods, groceries and mobile store business	Dis-continued business	-	Wholesaler market in Iwaki	-	Discontinued business due to the owner's age (75 years old)

Source: Based on authors' interviews.

households were involved, but this increased to 40 households and 68 individuals as of February 2013.[8]

Second, at the site of locally owned convenience store D, which had suspended operations in August 2012 (Table 9.2), a large convenience-store chain opened its Kawauchi Village branch store in December 2012.[9] This store carries 2,700 different items, making its selection of goods equal to, and perhaps even better than, that of a regular convenience store. The village has allowed the convenience store to operate rent-free for a period of three years. The store is crowded with local residents as well as workers involved with decontamination operations. Neither the Co-op nor the convenience store had been in operation in Kawauchi Village before the earthquake; both represent measures aimed at aiding the process of returning residents to the formerly evacuated areas.

Third, for their part, local retailers in Kawauchi Village have joined together through the framework of the local Chamber of Commerce and Industry to pursue the creation of a joint retail store. As a first step, they developed a joint ordering system in April 2013 with support from the national and prefectural governments. This system provides a collective means of purchasing products from wholesalers for interested merchants in the village who have lost their trading partners and must now travel far and wide to obtain products. Additionally, tablets have been distributed to individuals unable to shop easily for themselves, and an experimental system of using video conference calls to purchase goods has been introduced. This is a highly unusual service for a local retailer to offer.[10] As a transitional service until the opening of the joint retail store, the Co-op started a "traveling store" stocked with daily products, using a mobile sales vehicle.[11] With these initiatives, efforts, and support from residents, government offices, the Co-op, and local and non-local business organizations, commercial and shopping opportunities in the village are being restored, and in some cases have improved upon pre-disaster conditions.

Furthermore, there are signs of positive results in the reconstruction of the wider regional system of commerce in which Kawauchi Village is embedded. Following review of the "restricted area" designations (which meant an across-the-board restriction of entry to the town), on March 25, 2013 Tomioka Town was re-designated into "difficult to return," "limited residence," and "evacuation lift preparation" areas. Since the road connecting Kawauchi Village to Iwaki City (i.e. the Prefectural Ono-Tomioka Road and the National Route 6 to Iwaki City) was re-designated into "limited residence" and "evacuation lift preparation" areas, it became possible to travel from Kawauchi Village to Iwaki City via Tomioka Town.[12] The many truncated dead-ends found on the roads leading into Kawauchi Village after the earthquake have thus been eliminated. However, Tomioka Town's social and commercial capacities remain disabled. For this reason, the effects of Tomioka's re-districting on Kawauchi Village's commercial functions are not yet evident and can only be monitored in the future.

Discussion: building back better in Kawauchi Village?

We have focused here on efforts to restore commercial services and capacities in Kawauchi Village amid the extraordinary situation of a long-term all-resident evacuation followed by a slow and gradual return of residents. Connections between Kawauchi Village and surrounding towns and villages were severed by the nuclear disaster. Efforts have begun to secure commercial functions in order to support residents as they return to the village. Since commercial services are tightly connected to the daily lives of villagers, it is particularly important to restore these services in the near term. The process of restoring commercial services and capacities in Kawauchi Village has been aided by such activities as utilization of an existing but out-of-operation retail site for a new national-chain convenience store, and through the collaboration with the Japanese Consumers' Co-operative Union. As a result of these efforts, the quality of commercial services and capacities within the village has now been restored to, or even surpassed, the pre-disaster level. In addition, since the Co-op joint provisioning service and the mobile sales unit were not in existence prior to the disaster, they can be seen as examples of building back better.

Nevertheless, it is important to recognize contextual specificities that facilitated the rebuilding of commercial capacity in Kawauchi Village, lessons learned, and emergent problems, which in turn can inform future studies of BBB. First, the highly proactive policies to support the opening and reopening of convenience stores in Kawauchi Village were largely due to the fact that this was the first evacuated village to issue a "Declaration to Return." Since the national government is determined to see evacuees return to their former communities, large sums of money were invested in this test case. The establishment of the village's first national-chain convenience store encouraged residents to expect services equivalent to those of urban areas. Second, the activities undertaken by the Co-op were the result of villagers' proactive efforts to press for a more livable community. Restoration of commercial functions in Kawauchi Village was thus both a top-down and bottom-up process. Third, while shopping options have expanded for consumers in the village, local retail stores operating before the disaster are now forced into difficult competition with newly arrived convenience stores and the Co-op. There is potential for the already weak commercial base of existing retailers to encounter even more difficult conditions in the near future.

Finally, although we do not wish to trivialize the BBB efforts described in this chapter, we cannot sufficiently emphasize just how daunting the impairment brought about by the nuclear disaster is. The regional network of interlinked commercial services that connected Kawauchi Village to other communities such as Tomioka Town has only been partially restored. Although restricted areas have been re-designated, complete restoration of the various functions that previously existed in the coastal towns of Futaba Country will take considerable time (if it ever happens). The restoration of connections to inland cities such as Tamura, Miharu, and Koriyama has begun to become visible on the horizon, and broader regional efforts will be required to push for the restoration of Kawauchi Village's commercial and other functions in the future.

Conclusions

In this chapter we have analyzed efforts to restore and improve commercial functions and capacity in a locality that was compulsorily evacuated after TEPCO's nuclear accident and to which residents have begun to return, drawing on the Build Back Better (BBB) concept. We summarize the empirical findings of our study of Kawauchi Village as follows. First, although the pre- and post-disaster population of the village remains largely unchanged (at least in the official data), the number of households has increased. Such multiplication of households represents a characteristic type of damage in the case of a large-scale disaster leading to long-term evacuation of a rural area. One reason for this phenomenon is the multi-generational structure of households in agricultural districts in Japan. Following the disaster, entire households were not able to live together in small temporary housing, and many families were subsequently divided along generational lines. Another reason for such separation of households is related to the specific effects of a nuclear disaster. In many cases, mothers took their children into evacuation at far distant sites in order to shield them from the effects of radiation (Seto *et al.* 2015), while fathers stayed behind to be close to employment sites. This was one of the major sources of household separation. It should be noted that prevention of household separation is linked with the speed of population recovery after residents return. To prevent separation of households, it is imperative to provide temporary housing tailored to the specific needs of the area, ensure the safety and security of children and other vulnerable individuals, and, finally, improve livability, particularly through restoration of commercial capacities. A frequently voiced opinion is that it will be highly difficult for villagers who have enjoyed the amenities of urban life to return to rural villages (although see Kaneko 2016). However, proactive restoration of commercial capacities represents one means of lowering the barriers for evacuees to return after temporary housing rights have expired. Improved commercial capacities are also highly important for attracting new residents.

Second, to restore the commercial capacities of the regional distribution network, active appeals from local people in the afflicted areas are critical, in addition to various forms of "outside" support. In the case of Kawauchi Village, action was not only top-down from the government but also bottom-up, and grassroots lobbying and initiatives were linked to improvements in livability. The key players in the process included consumer co-operatives, which have a unique institutional history in Japan; taking advantage of this organizational resource has been critical in Kawauchi Village. Additionally, the private sector, such as large convenience-store chains, played an important role not only in speedy recovery, but also in effectively raising residents' consciousness of reconstruction. In short, we have observed several key social and organizational innovations to overcome the challenges associated with BBB articulated by Mannakkara and Wilkinson (2014).

Third, despite evidence of building back better, we must also point out emergent challenges. Especially when viewed from the perspective of existing retailers, post-disaster support tends to be biased towards new business operators. This phenomenon is not limited to retail but is seen across sectors. The provision of robust support for pre-existing businesses to continue operations after the disaster is an important point to consider further when thinking about how to build back better in disaster-afflicted areas.

Finally, in order for local communities to recover, it is imperative for evacuees to return and to establish conditions wherein various social activities can be resumed. A wide region was affected by the nuclear disaster, as was an economic network interlinking multiple municipalities. While recovery and restoration have focused on municipalities as discrete communities, there is a real need to restore and rebuild the entire economic network of the region, as was noted following the Canterbury earthquake in New Zealand (CERA, 2013). The areas affected by the nuclear disaster extend across a wide regional network not bounded by the borders of any single municipality. This regional network had important commercial functions, but its importance is not limited to commerce. The regional network system also played an important role in structuring local sociality and identity. Accordingly, even when evacuation comes to an end for an individual community, if the regional economic network is not fully restored, restoration from the disaster remains incomplete.

Acknowledgments

We would like to thank those who cooperated with our interviews. This study was supported by JSPS KAKENHI Grant Number 25220403 (Grant-in-Aid for Scientific Research (S), 2013–2017, Principal investigator: Mitsuo Yamakawa).

Notes

1 From documents of the Fukushima Disaster Response Center. Gray (Gy) represents the amount of radiation energy applied per unit mass of an object. One gray implies that 1 joule (a unit for measuring energy) of energy was absorbed by 1 kg of the substance. Sievert (Sv) is a unit for representing the health effects of radiation when humans are exposed (www.aomori-hb.jp/ahb3_5_6_06.html; last accessed August 9, 2015). Also see Yamakawa and Yamamoto (2016, 8–11).

2 When evacuees from Tomioka Town and Kawauchi Village were here from March 16 to August 11, 2011, it was the largest evacuation facility in the prefecture. Up to 2,500 evacuees were housed here at one time (The Publication Committee of Alive, Living, and Life 2011).

3 The Basic Resident Register is maintained in accordance with the Basic Resident Register Act and records information pertaining to residents of municipalities.

4 From internal documents of Kawauchi Village.

5 From surveys conducted in Kawauchi Village in June 2011.

6 From the Fourth Comprehensive Plan of Kawauchi Village (www.kawauchimura.jp/outline/synthesis_plan.html; last accessed August 15, 2015).

7 "Newsletter of Co-op Action for Connecting" (February 27, 2013).

8 The Japanese Consumers' Co-operative Union (Co-op) is one consumer cooperative established under the Consumer Cooperatives Act. Consumers become members by paying dues and cooperatively manage the organization. One of the main activities is collective purchase operation, in which they jointly purchase fresh foods and daily necessities, and receive home delivery (http://jccu.coop/eng/; last accessed July 13, 2015).

9 Family Mart homepage (www.family.co.jp/company/news_releases/2012/121130_3.html; last accessed July 7, 2013).

10 From documents from Fukushima Prefecture, Kawauchi Village, Kawauchi Village Commercial Association, and others. In 2012 the authors participated in the "Project to Support Shopping by Utilizing ICT" as directors.

11 "Newsletter of Co-op Action for Connecting" (May 16, 2013) (http://shinsai.jccu.coop/shien/2013/05/post-141.html; last accessed July 16, 2015).

12 Following this, and in line with the progress in decontamination, improvements were made to the road network. On September 15, 2014, all lanes of National Road 6 from Tomioka Town north to Namie Town were reopened (www.kahoku.co.jp/tohokunews/201409/20140915_63018.html; last accessed July 17, 2015). Later, on March 1, 2015, the Joban Expressway from Joban Tomioka interchange to Namie interchange was reopened, with all lanes passable (http://jobando.jp; last accessed July 17, 2015). With all the routes linking the region of Hama-Dori south to north reopened, the efficiency of distribution was expected to improve.

References

Basic Resident Registers. 2012 and 2013. e-Stat: Portal Site of Official Statistics of Japan. Accessed December 21, 2015. www.e-stat.go.jp/.

Census of Japan. 2012. e-Stat: Portal Site of Official Statistics of Japan. Accessed December 21, 2015. www.e-stat.go.jp/.

CERA. 2013. "Canterbury Earthquake Recovery Authority." Accessed July 13, 2015. http://cera.govt.nz/2013.

Clinton, W. J. 2006. *Lessons Learned from Tsunami Recovery: Key Propositions for Building Back Better*. New York: Office of the UN Secretary-General's Special Envoy for Tsunami Recovery.

Fan, Lilianne. 2013. *Disaster as Opportunity? Building Back Better in Aceh, Myanmar and Haiti*. London: Humanitarian Policy Group Overseas Development Institute.

Kaneko, Hiroyuki. 2016. "Radioactive Contamination of Forest Commons: Impairment of Minor Subsistence Practices as an Overlooked Obstacle to Recovery in the Evacuated Areas." In *Unravelling the Fukushima Disaster*, edited by Mitsuo Yamakawa and Daisaku Yamamoto, pp. 136–153. London: Routledge.

Kennedy, J., J. Ashmore, E. Babister, and I. Kelman. 2008. "The Meaning of 'Build Back Better': Evidence From Post-Tsunami Aceh and Sri Lanka." *Journal of Contingencies and Crisis Management* 16(1): 24–36.

Khasalamwa, S. 2009. "Is 'Build Back Better' A Response to Vulnerability? Analysis of the Post-Tsunami Humanitarian Interventions in Sri Lanka." *Norwegian Journal of Geography* 63(1): 73–88.

Lyons, Michal. 2009. "Building Back Better: The Large-Scale Impact of Small-Scale Approaches to Reconstruction." *World Development* 37(2): 385–398.

Mannakkara, Sandeeka, and Suzanne Wilkinson. 2014. "Re-Conceptualising 'Building Back Better' to Improve Post-Disaster Recovery." *International Journal of Managing Projects in Business* 7(3): 327–341.

Seto, Masayuki, and Akira Takagi. 2014. "Comparison of the Disasters and Spatio-temporal Scale." *The 9th China-Japan-Korea Joint Conference of Geography*, Proceedings, 37–74.

Seto, Masayuki, Akira Takagi, Tamaki Honda, and Rie Imaizumi. 2015. "Development of Disaster Reconstruction Model and Application to Post-Disaster Education Environment." *The 10th China-Japan-Korea Joint Conference of Geography*, Proceedings, 48–52.

The Publication Committee of Alive, Living, and Life. 2011. *Alive, Living, and Life: The 169 Days at the Fukushima Big Palette Evacuation Shelter 2011*. Tokyo: UM Promotion.

Yamakawa, Mitsuo, and Daisaku Yamamoto (eds). 2016. *Unravelling the Fukushima Disaster*. London: Routledge.

10 Renewable-energy policies and economic revitalization in Fukushima

Issues and prospects

Yoshio Ohira

Introduction

The Great East Japan Earthquake Disaster reaffirmed the vital importance of electricity. Immediately after the earthquake, more than eight million households lost power as the result of extensive damage to power plants, substations, and power lines caused by the earthquake and tsunami. Efforts to bring these plants back online were initially ineffective. Accordingly, power-generation capacity was unable to meet demand, leading to the implementation of rolling blackouts in the Tokyo metropolitan area from March 14 to March 28. The nuclear accident at the TEPCO Fukushima Daiichi Nuclear Power Plant (NPP) also raised concerns about the safety of all nuclear power plants, leading to a nationwide suspension of operations that, in turn, forced a number of out-of-operation thermal power plants to be brought back online to make up for the energy-supply deficit. According to the Agency for Natural Resources and Energy (2013), greatly increased imports of natural gas and oil are estimated to have increased fuel costs for power generation by 2.3 trillion yen in 2011 and 3.1 trillion yen in 2012. Since Japan imports nearly all of its fossil fuels, these increased expenditures have serious consequences for the nation's balance of trade.

In this context, it is not surprising that renewable-energy sources such as solar and wind have become a topic of great interest and expectation in post-Fukushima Japan, or that a shift in renewable-energy policy has been pursued. Prior to the disaster, national renewable-energy policies had been largely ineffective (e.g. Oshima 2010). Following the disaster, aggressive policies were implemented to rapidly increase the output of renewable energy, particularly in Fukushima Prefecture. In addition to the obvious material consequences of the nuclear disaster for the prefecture's important energy economy, Fukushima was stigmatized by the disaster in less obvious but no less consequential ways. "Clean," renewable energy has thus been aggressively pursued both as a strategic economic and a public-relations measure. The results have been staggering. For example, more than 40,000 solar energy-generation facilities have been proposed in the prefecture since the disaster, and more than half of the facilities are already in operation. Clearly, many in the prefecture are placing their bets on renewable energy as the pathway to revitalization.

While the image-enhancing side effects are important, it is clear that renewable-energy industries have been promoted first and foremost as a means of generating employment and revitalizing local economies. Here, at least three distinct motivations can be identified. First, the hopes placed on renewable energy should be seen as a response to another round of industrial shifts brought on by technological and political economic change at the global scale. Sericulture and coal were once prominent industries in Fukushima Prefecture. However, following the Great Depression and the development of synthetic fabrics, sericulture was largely abandoned in favor of other forms of agriculture, particularly fruit orchards. Similarly, the coal industry in the prefecture was devastated by the switch to oil and eventually replaced as a leading industrial sector by manufacturing and tourism. These key industries now face increasing global competition, liberalization of trade (e.g. political initiatives for the Trans-Pacific Partnership), and a shift away from greenhouse gas-producing energy. The second motivation stems from damage to regional industries resulting from the TEPCO nuclear accident, including the highly negative effects of radioactive contamination for the region's agriculture and tourism. Despite the eventual implementation of a rigorous food-inspection system, local agriculture continues to suffer from the stigma attached to the prefecture (Chapter 6). Tourism has also seriously suffered: annual tourist visits in the prefecture dropped from 57 million in 2010 to 35 million in 2011. In 2013, annual tourist visits recovered to 48 million, but they stayed around 47 million in 2014, significantly below the pre-disaster level (Fukushima Prefecture n.d.). A third motivation arises from the strong dependence of many coastal localities on the jobs, tax revenues, and subsidies provided directly and indirectly by the nuclear power plants (i.e. Fukushima Daiichi and Daini). As is widely known, this structural dependence was created through various policies, such as the "Three Power Siting Laws," implemented from the postwar period forward (Oshima 2013; Kingston 2013; Samuels 2013). In 2009, the electricity, water, and gas industries accounted for more than 65 per cent of the economy in the four municipalities in which Fukushima Daiichi is situated (i.e. Nahara, Tomioka, Okuma, and Futaba Towns), but in 2011 these same industries accounted for only 0.5 per cent of each local economy. These municipalities used to record some of the highest per capita incomes in the prefecture, but now have some of the lowest. The nuclear disaster forced people in Fukushima to critically re-examine the economic and fiscal dependence orchestrated by the "nuclear village" in exchange for energy provision for the Tokyo metropolitan area. Rebuilding the economic bases of these municipalities, without reproducing the structural dependence, is clearly an urgent task.

Taken together, these motivations explain why renewable energy is seen as a potential key industry in Fukushima today. Yet even the best of intentions do not always produce beneficial results. The expansion of renewable energy does not necessarily entail regional economic development. A case in point is the system of Feed-in Tariffs (FITs) that was initiated in Japan in July 2012. The

FIT system establishes the purchase price and purchasing period of renewable energy beforehand and thus makes it easier to judge the merits of investing in renewable-energy projects. The introduction of this system accelerated the implementation of large-scale projects, including most prominently so-called "mega-solar" projects, throughout the country. Renewable energy-generation projects are currently increasing within Fukushima, to the extent that this prefecture is now the leader in certified installations. However, there are many cases where these projects' benefits for the people of the prefecture are virtually nil (*Fukushima Minyu*, 2014). Hence there is a growing suspicion that increasing renewable-energy generation will automatically bring positive local results without conscious efforts to link renewable energy to employment generation and the revitalization of local economies.

Existing studies on renewable-energy development rarely address the issue of linking renewable energy with local economic development. Most studies instead focus on the effectiveness of policies for promoting renewable-energy projects (e.g. Chowdhury *et al.* 2014; Menanteau *et al.* 2003; IEA 2008; Oshima 2013). Menanteau *et al.* (2003), for example, argue that the effects of FIT policies are similar to those of competitive price-bidding systems such as the Renewable Portfolio Standards (RPS) system, but that FITs are more efficient due to their ability to secure stable investments.[1] The IEA (2008) also maintains that FIT policies can secure low risks and incentives for power sources in need of market competitiveness in the initial stages of development, and are effective for rapidly expanding renewable energy as an infant industry. Alireza and Mohaghar (2013) offer a somewhat different perspective by identifying unique aspects of the renewable-energy industry, but their underlying motivation remains the promotion of the industry itself. These studies, in other words, are essentially concerned with the question of how to promote and expand renewable energy, and rarely question whether and how renewable-energy industries can benefit localities in which renewable-energy facilities are actually built and operated.

With the above considerations in mind, this chapter examines how renewable energy can be implemented to promote employment generation and the revitalization of local economies. In order to do so, I first compare renewable-energy policies in Japan and Germany. Germany is a leader in the field of renewable-energy policies and the German experience with renewable-energy policies has had considerable influence on FIT systems in Japan, and particularly in Fukushima Prefecture. It is thus imperative to use the German case to proactively identify challenges involved with FIT policies before they develop in Japan. We then turn our attention to the locally oriented renewable-energy initiatives that emerged after the nuclear disaster, identifying their successes, challenges, and prospects. Finally, I outline a system of *chisan-chisho* (i.e. local production for local consumption) in which energy produced in Fukushima Prefecture is consumed in the prefecture. We will explore whether such a system of energy independence is feasible, and what shape it could potentially take.

Renewable-energy policies

Renewable-energy policies in Japan

FIT policies were first introduced in Japan in July 2012.[2] The FIT system requires electrical-power companies to purchase renewable energy (i.e. energy produced through solar, wind, small-scale hydro, thermal, and biomass power generation) at a fixed price for a specific period (i.e. the "purchasing period"). The fixed price and purchase period for renewable energy under this system vary according to the generating source. For example, the fixed price for >10 kW solar energy—the most frequently introduced source of renewable energy—has been lowered from 40 yen/kWh (kilowatt hour) in 2012 to 36 yen/kWh in 2013 and 32 yen/kWh in 2014. The purchase period for >10 kW solar energy has been set at 20 years. The fixed price and purchase period for other sources of renewable energy have remained stationary since being introduced in 2012. As evidenced by the extension of FIT to offshore wind generation and existing hydropower beginning in fiscal year 2014, the FIT policy system is being continually revised and amended.

The increased costs of purchasing renewable energy under the FIT system are passed along to consumers in the form of a renewable-energy surcharge added onto electricity bills. Thus it is consumers who foot the FIT bill. As the proportion of renewable energy in the overall energy mix increases, electrical-power companies raise their fees in order to offset the higher costs of renewable energy. With the increasing proportion of renewable energy in the overall energy mix, the initial surcharge fee of 0.22 yen/kWh in the inaugural year of 2012 rose to 0.35 yen/kWh in 2013 and 0.75 yen/kWh in 2014. Although data is only available for three years in Japan, a comparison of surcharge fees for the first three years of FIT implementation shows a more rapid increase of fees in Japan than in Germany. Accordingly, it is imperative to adopt measures to somehow curb the burden of rising surcharges.

While Germany is a leader in the field of FIT systems and policies, the rising burden of surcharges remains an issue there as well, and FIT policies are periodically revised in light of this issue. Thus a comparison of FIT policies in Germany and Japan is highly instructive.

Renewable-energy policies in Germany and Japan

In Germany, an "Electricity Feed-in Law" (*Stromeinspeisungsgesetz*) was enacted in 1991 that required energy companies to purchase renewable energy at a fixed ratio based on average energy fees, and in 2000 the passage of the Renewable Energy Purchase Law (*EEG: Erneuerbare-Energien-Gesetz*) required energy companies to purchase renewable energy at a fixed purchase price. Following the introduction of EEG, nearly 800 energy cooperatives composed of nearly 200,000 citizens, businesses, and agriculturalists were established. These energy cooperatives have initiated numerous renewable-energy projects and have made a considerable contribution to the expansion of renewable energy.[3] The EEG was amended in 2004, in 2008, and again in 2011 and, among other changes, the fixed price was amended. Due in part to these revisions, renewable energy has rapidly increased and, in 2004, Germany overtook Japan as the world leader for installed capacity of solar energy. Chowdhury *et al.* (2014) identify the security of long-term revenue streams from solar energy-generation projects as one of the reasons for the expansion of this sector in Germany.[4]

While the German EEG is a successful and leading example, the rising cost of renewable-energy surcharges added onto electricity bills remains an issue. As indicated in Table 10.1, surcharges arising from EEG were 0.2 eurocent/kWh in 2000 but rose to 3.5 eurocent/kWh in 2011. In order to respond to the rapid increase in renewable energy throughout Germany, the EEG was revised in 2008 and a system of total volume control was established. The addition of total volume control means that when a pre-determined amount of renewable energy is reached, the fixed price will decrease.

In Japan, attempts have been made to control the rising cost of surcharges by lowering fixed prices on an annual basis. The problem is, however, that lowering the fixed price negatively impacts the expansion of renewable-energy generation. Particularly in Fukushima Prefecture, where there is a need to expand renewable energy over the long term, it is necessary to adopt measures to counteract annual reductions of fixed prices. With this in mind, the next section discusses frameworks that can continue to support the introduction and expansion of renewable energy under the FIT system.

Renewable-energy policies in Fukushima and issues for the future

Renewable-energy policies in Fukushima

In March 2013, the "Fukushima Renewable Energy Promotion Vision (revised edition)" (REvision) was announced. This plan aims, by 2040, to generate more renewable energy in the prefecture than the total amount of energy consumed there, and also establishes the goal of making Fukushima Prefecture a leader in the field of renewable energy. It aims to construct frameworks through which citizens will become key agents in energy policy, funds are circulated locally, the benefits of renewable-energy development are realized locally, and

Table 10.1 Changes in FIT surcharge based on the German Renewable Energy Purchase Law (EEG: Erneuerbare-Energien-Gesetz)

	2000	*2002*	*2004*	*2005*	*2006*	*2007*	*2008*	*2009*	*2010*	*2011*	*2012*
EEG	0.2	0.3	0.5	0.6	0.8	1.0	1.1	1.3	2.1	3.5	3.31

Unit: Eurocent/kWh

Source: Federal Ministry for the Environment, Nature Conservation and Nuclear Safety (BMU) (2011, 2012, 2013).

renewable-energy systems based on local production for local consumption can be realized. Furthermore, through the advance of world-leading renewable-energy projects, such as experimental studies of offshore floating wind turbines, the prefecture is aiming to develop a high concentration of renewable energy-related industries and employment.

In terms of more concrete measures and projects guided by the REvision, three initiatives are noteworthy. First, in February 2013 Fukushima Prefecture established the Fukushima Renewable Energy-sector Network (FRENET), an organization that aims to support renewable-energy generation by building a network of renewable-energy operators.[5] While renewable-energy projects have previously been implemented primarily by energy companies, and occasionally by private businesses and local governments, the new FIT policies have made it much easier for a wide range of actors to easily initiate projects. However, there is a lack of know-how regarding renewable-energy projects and many cases in which actors are unsure of how to plan renewable-energy operations. Another source of difficulty stems from the fact that renewable-energy projects remain few and it can be difficult to secure funding. To overcome these issues, FRENET is providing support for renewable energy-project planning and sharing the challenges of renewable-energy projects through networking. Second, the Fukushima Electric Power Company was established in 2013, initially through funds from the prefectural government and later through funding from municipal governments and private companies. The company used a citizen-participation fund, in addition to conventional funding from financial institutions, to implement a mega-solar energy project at Fukushima Airport. It uses the profits from the sales of electric power and accumulated business experience and knowledge to provide lectures on technical and business-management aspects, and more general information, to parties interested in renewable-energy projects. Third, the prefectural government also offers subsidies for feasibility studies for introducing hydro and wind-power generation in order to reduce uncertainty associated with the profitability of renewable-energy projects and other business risks.

In summary, these organizations and policy measures provide a vehicle for addressing the many challenges that arise during the process of implementing renewable-energy projects on the ground, and critically complement the stimulus provided by the FIT system. The increase in renewable-energy projects in Fukushima should be attributed at least in part to these local-level initiatives and policy efforts.

Renewable-energy generation in Fukushima: current state and future issues

It is necessary for renewable energy-generation projects implemented in Fukushima Prefecture to be linked with efforts to recover from the earthquake and nuclear disaster. The goal is not simply to introduce or increase renewable energy generation, but rather to link renewable-energy projects with the revitalization of the local economy. Figure 10.1 presents data pertaining to changes in the amount of solar-energy power generation approved under FIT policies in Fukushima Prefecture following the introduction of FIT in Japan, as well as changes in the actual rate of implementation of these certified projects. Solar-generation devices of <10 kW show very high operating rates because most are photovoltaic panels which can be placed on roofs of residential homes, and are thus relatively easily and quickly installed. For-profit solar power-generation operations of >10 kW are typically implemented by power companies. Securing profits entails efforts to capture economies of scale, and thus renewable-energy operators tend to develop large-scale operations in areas where land is inexpensive. Accordingly, the amount of certified output from "mega solar" projects (i.e. solar energy with output over 1,000 kW) in Fukushima Prefecture is quite large. Partially due to the aggressive renewable-energy policies in place, as of December 2014 Fukushima Prefecture had become the top prefecture in Japan for the amount of certified renewable energy generated. Nevertheless, the actual operating rate remains low for approved projects of >10 kWh (Figure 10.2). Most of the "mega-solar" projects are not yet in operation, but rather are under construction or awaiting construction. This is due to the expected decline in the price of solar panels as well as the delayed decontamination process.

In order to implement renewable-energy projects that benefit Fukushima Prefecture, it is imperative that businesses and the people of the prefecture become

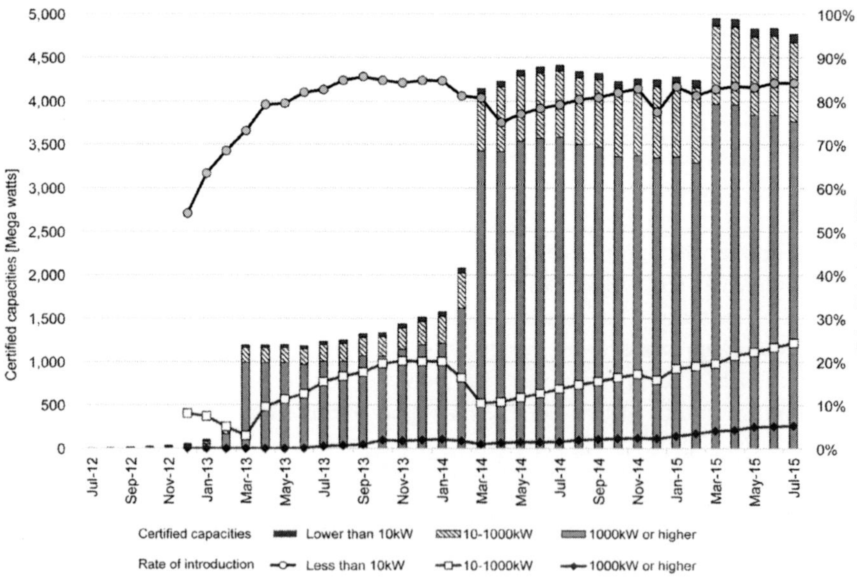

Figure 10.1 Changes in certified capacity and operation rate for solar-power generation in Fukushima Prefecture.

Note: Created by the author based on the Agency of Natural Resources and Energy (n.d.). While increases in the amount of certified output can be seen for March 2013 and March 2014, this is the result of the annual decrease in fixed prices under the FIT system leading to surges in demand.

the main agents of these projects. Less than 20 per cent of the solar energy-generation projects planned or in operation in Fukushima Prefecture are being implemented by companies based in the prefecture (*Fukushima Minyu*, 2014). Additionally, of the 78 mega-solar projects that have been publicly announced, 35 are being implemented by companies that do not have main offices, branch offices, or production facilities in the prefecture. Looking at the total amount of electrical-generation output, including planned output, 77.6 per cent of electricity is to be generated by companies from outside the prefecture. Thus, while profits will accrue to the companies behind renewable energy-generation projects, these profits will not stay within Fukushima. Expansion of solar energy-generation projects will create temporary employment in construction, periodic maintenance jobs, and some increases in property-tax revenues, yet these jobs will not make the large or sustained contributions to employment generation and revitalization of local economies that are needed to achieve recovery from the earthquake and nuclear disasters. If companies implementing renewable energy-generation projects are based in Fukushima Prefecture, then the profits from these projects will flow into employee incomes and can be expected to stimulate and expand local economic activity.

Because renewable-energy projects, including solar energy, are, generally speaking, rather weak generators of employment, their ability to benefit local economies has been questioned. One of the exceptions is woody biomass energy, which creates local employment opportunities in forestry and lumbering operations to procure the wood chip to be combusted, and in which power generation itself requires a regular workforce. Fukushima has strong potential for biomass energy given the large areas of the prefecture covered by forests, and in fact woody biomass power is the next largest type of renewable energy after solar power in terms of quantity certified (Agency for Natural Resources and Energy 2013). Yet, the prospects for biomass energy in Fukushima were clouded by the nuclear accident. Forests and forest products were contaminated as a result of the nuclear accident, and many people fear that radioactive materials will be emitted through biomass incineration. As a result, an opposition movement has formed, creating obstacles to the implementation of woody biomass power-generation projects. Although research indicates that filters can remove 99.99 per cent of all radioactive materials (National Institute for Environmental Studies 2014), the advance of woody biomass power generation has been slowed considerably: an additional reason for the focus on revitalization of local economies through solar power as an important strategy.

Proposals and prospects for renewable energy in Fukushima Prefecture

Local revitalization through collaboration with local businesses

Until now, research on renewable-energy projects has focused on renewable-energy policies such as FIT, and research on individual renewable-energy projects has focused almost entirely on technological development. Research focusing on the implementation of renewable-energy projects by local companies has been rather scarce. Here I examine the Shirakawa Regional Renewable Energy Promotion Council (SRREPC) as an example of a local private-sector initiative attempting to link renewable energy-generation projects to employment creation and the revitalization of local economies. The case study highlights the importance of involving local companies and individuals as active agents and of taking advantage of their assets and resources to overcome obstacles to renewable-energy projects such as securing funds, barriers to information, and cost competitiveness.

The SRREPC was formed through the banding together of members of a local voluntary economic association from the southern part of Fukushima Prefecture—the Group of Small and Medium Firm Entrepreneurs—around the goal of overcoming the issue of stigma (*fuhyo higai*, or "reputational damage") that has plagued the prefecture since the nuclear accident, through the promotion of renewable energy within the prefecture. Major activities of the council include conducting research related to renewable energy, and information dissemination. Members of the SRREPC have led the promotion of solar energy-generation projects of >50 kWh. Thus far projects totaling over 2 MW (megawatts) have been implemented, primarily in Shirakawa City. One key characteristic of the SRREPC is that its individual members are involved with the management of various businesses (their "main businesses")—including real estate, construction, and electrical contracting—in the southern part of the prefecture.

There are three major benefits to local business owners and managers being involved in renewable-energy projects as "auxiliary businesses." First, solar-power generation is, from a technical standpoint, rather simple to plan and develop, because the viability of solar-power generation itself is in large part determined by well-defined physical properties such as the amount of available sunlight and topography. Accordingly, individuals who have local business expertise can determine the feasibility of solar-energy projects relatively easily as long as they have access to such baseline knowledge.

Second, securing funding becomes easier when local business operators are involved, for even if procuring funding is difficult based on the operational plan for the solar energy-generation project alone, trust and reliability are increased by knowledge of the SRREPC members' main business activities, making decisions to finance easier for financial institutions.

Third, local businesses and renewable-energy (i.e. solar-energy) projects can mutually benefit each other, and this collaboration can also help support disaster recovery. What is meant here by collaboration is the advantageous utilization of local industrial expertise (i.e. knowledge and technology) to implement renewable-energy projects that lead to reduced costs for renewable-energy projects or reduced costs and increased income for local businesses. For example, local real-estate agents are knowledgeable about laws and regulations on mortgages and land use, as well as the local availability of land. Thus their involvement in solar energy-generation projects enables smooth implementation in accordance with various local land-use regulations.[6] Local construction companies and contractors can take advantage of their knowledge and construction skills and their supply channels of materials to build low-cost wood framings for solar panels. In this way, local business owners' and entrepreneurs' participation in renewable-energy projects can reduce the costs of these projects, and power generation can be made more profitable.

The SRREPC capitalizes on these advantages of collaborative efforts among local entrepreneurs to promote local renewable-energy projects. This brief case study helps us draw several implications. First, since the types of local businesses represented in SRREPC membership include construction, real estate, and other ubiquitous business forms that exist throughout the country, the SRREPC can potentially serve as a model that is applicable in other regions. Second, recall that the original goals of the SRREPC were to combat stigma and to revitalize the local economy. Renewable energy is conceptualized as a means of reaching these aims. Despite the positive early signs in terms of promoting renewable-energy projects and anecdotal stories of successful business operations, objective assessment as to whether and to what extent the council has been successful in achieving the two primary goals remains incomplete. Last, while in Germany renewable-energy projects have increased through the efforts of civic cooperative associations, in Japan such examples are rare, with firms remaining the main agents behind renewable-energy projects. Table 10.2 provides information related to the main agents behind solar-power generation projects in Japan. Corporations account for 90.8 per cent, while solar power-generation projects implemented by cooperative associations total only 3.2 per cent. The above case study implies that there may be cooperative *business associations* like the SPREPC behind individual firms, which facilitate information exchange and coordinate actions among a diverse range of companies with their main business outside of renewable energy. Future studies may examine the comparative advantages and disadvantages of civic and business associations in facilitating renewable-energy development.

Energy policies to expand renewable energy over the long term and promote local production for local consumption

While the active involvement of local businesses is essential in promoting renewable energy that also benefits local economies, the role of local governments

Table 10.2 Approved solar power-generation projects and implementing agencies in Japan

	Number of projects	*%*
Company, limited company, limited liability company, and partnership in commendam	984	90.8
Cooperatives and agricultural corporation	35	3.2
Local government	19	1.8
Other	46	4.2

Note: Since data on the agencies behind renewable-energy projects implemented under FIT are not publicly available, data from Japan Photovoltaic Energy Association, Reconstruction Center (JPReC) (various years) on subsidies for approved projects are used as a surrogate.

remains vital. Fukushima Prefecture has proposed the pursuit of an energy system based on the principle of *chisan-chisho* (local production for local consumption) through the promotion of renewable energy. That is, it aims to produce energy primarily for local consumers within the prefecture. This stance is evidently the prefectural government's response to the nuclear accident and underlying structural problems exposed through the accident. Here the distance between an energy-supplying region, such as Fukushima, and an energy-consuming region, such as the Tokyo metropolitan area, overlaps with the asymmetrical power relations between the nuclear host community and the "nuclear village." In this way, *chisan-chisho* is seen as a means to regain and retain decision-making power over the management of energy production by narrowing the gap between the sites of energy production and consumption. Demonstrating the potential for *chisan-chisho*, it is hoped, will help to create a model of a sustainable society reliant on renewable rather than nuclear energy.

The aspiration towards *chisan-chisho* is also rationalized by the presence of many natural environments suitable for various renewable-energy projects in Fukushima, so that efforts can be made to distribute the risks of unstable renewable-energy projects across wide areas of the prefecture. Therefore, increasing the volume of renewable energy in Fukushima would also mean increasing its diversity by being sensitive to locally specific conditions. Accordingly, it is imperative to advance renewable-energy operations not only through such local business organizations as the SRREPC, but also through the efforts of each local municipality in the prefecture, with foresight, resources, and sensitivity to unique natural environmental and socioeconomic conditions. In what follows I will attempt to outline such a vision.

Fukushima Prefecture is conventionally divided into seven different regions based on landforms and urban structure, and each region has distinct physical and built environmental characteristics. For example, while the relatively flat Iwaki region is well suited to solar energy-generation projects, the mountainous Aizu and Minami-Aizu regions are well suited to hydropower generation. More densely populated areas such as Kenpoku (in which Fukushima City, the prefectural capital, is located), Kenchu (around industrial Koriyama City), and Iwaki (around the commercial center of Iwaki City) are likely to show a renewable-energy deficit within the region (i.e. demand greater than supply), while other regions would likely show a renewable-energy surplus.

The unique characteristics of this environment can be utilized to conceptualize and construct a three-stage process for realizing an energy system based on the principle of *chisan-chisho* through renewable energy (see Figure 10.2). In the first stage, in each of the seven regions of Fukushima Prefecture (i.e. the seven circles in Figure 10.2), local people strive to understand the characteristics of the natural environments and industrial structures of their region and also analyze the state of regional energy demand—for example, whether the energy system of their region is currently characterized by excess demand, excess supply, or balance. Analysis and comprehension of the local situation makes it possible to evaluate the types and degree of renewable energy that can be introduced in the future. In the second stage, analysis of the local energy system can then be utilized to create flexibility between regions within the prefecture. Through the complementarity of excess supply and excess demand between regions (i.e. the arrows connecting regions in Figure 10.2), it will become possible to achieve local production for local consumption at the prefectural level. This second stage entails a number of challenges regarding inter-regional flexibility, including the advance of energy liberalization such that all consumers can freely select their energy provider, putting in place the infrastructure for energy transmission via smart grids, and disclosure of information pertaining to local energy operators such as amount generated and amount consumed. In the third and final stage, energy produced within Fukushima Prefecture is provided to areas beyond the prefecture (i.e. the vertical arrows in Figure 10.2).

There is a limit to the amount of energy demand within Fukushima Prefecture, and this limit sets possible constraints on the introduction of renewable energy

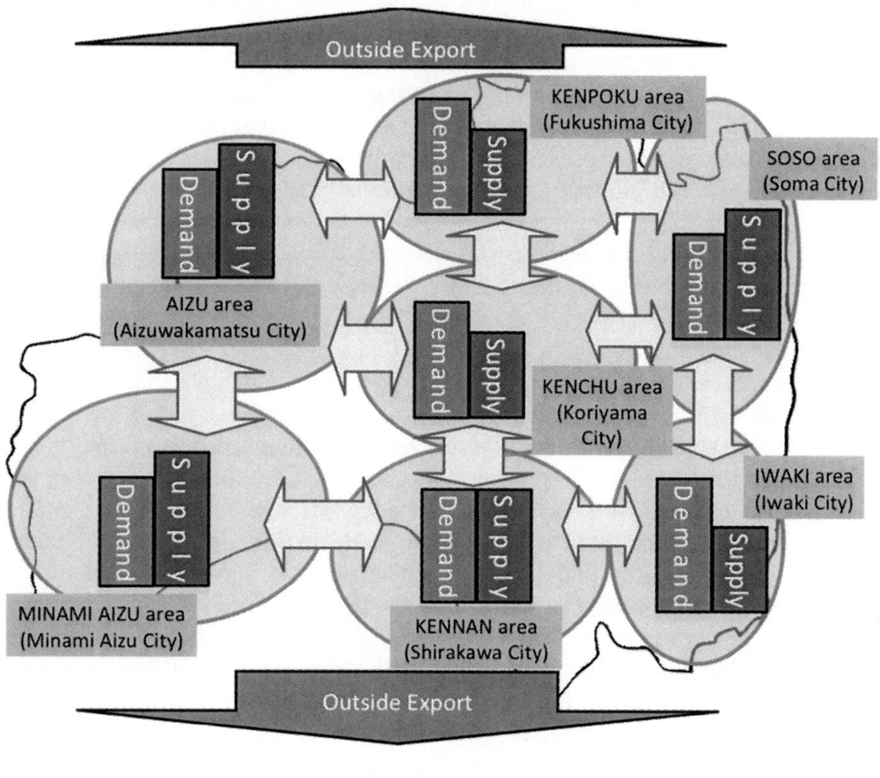

Figure 10.2 Schematic image of *chisan-chisho* (local production for local consumption) for energy in the seven regions of Fukushima Prefecture.

Note: Based on Ohira (2013). The seven ovals indicate cultural regions of the prefecture. The bar graphs indicate the balances of energy supply and demand, which are variable based on season and time. The arrows connecting these regions portray the transfer of surplus energy between regions in order to stabilize the overall energy supply.

within the prefecture, but the possibility of supplying energy to meet the abundant demand beyond the prefecture makes it possible to stabilize and expand Fukushima's renewable energy-generation projects. Stabilizing and expanding renewable energy-generation projects is important for two reasons. First, it is desirable from a business-management perspective. If the abundant energy demand outside the prefecture can be captured, then it becomes possible to reap profits from beyond Fukushima Prefecture. Second, stabilization and expansion enables a stable energy supply within the prefecture. Since the output from individual renewable-energy projects is small and often reliant on rather unreliable natural sources, stabilization of renewable-energy supply depends on increasing both the number and type of renewable energy-generation projects. If only energy demand within Fukushima Prefecture is targeted, then excess supply must be curbed in accordance with daily and seasonal fluctuations in demand, thus resulting in inefficiencies. Emphasizing and pursuing markets beyond the prefecture offers a solution to these issues.

Finally, it is imperative to point out the role of municipal governments in the expansion of renewable-energy generation. The seven regional zones described above correspond to the major municipal regions of Fukushima Prefecture, and since the smallest unit for implementing policy is the municipality, it is best for energy policy to be conceptualized and enacted at the municipal level. Within the three-stage process of advancing towards an energy system based on *chisan-chisho*, municipal governments must play a particularly important role in the first stage. The local municipality has the best information about the environment and urban industrial structure of each respective city region and is thus in the best position to design energy policies rooted in this local environment.

As already mentioned, Fukushima Prefecture established REvision in March 2012. For individual municipalities, it provides basic guidelines and overall targets on which their own renewable-energy plans can be based. By adapting to local environmental and socioeconomic conditions, these municipal-level plans should help expand the introduction of renewable-energy generation. Nevertheless, the development of effective plans requires detailed baseline surveys that examine the feasibility of renewable energy. Some municipalities already have taken important steps towards that end, conducting detailed surveys as part of the Local New Energy Visions Project by the New Energy and Industrial Technology Development Organization (NEDO), a national government-sponsored agency, between 1998 and 2010. However, such baseline surveys and Local New Energy Visions were not generated for all municipalities and thus there are areas where they will need to be created from scratch. Moreover, citizens' views of energy policy have been transformed as a result of the earthquake and nuclear disaster, and even in municipalities where Local New Energy Visions were created, there is a need to revise these visions in line with shifting conceptions of renewable energy after the disaster. In short, there is much groundwork to be done at the local municipal level; this is indispensable to realizing *chisan-chisho* based on renewable energy in Fukushima.

Conclusion

Historically, local economic-recovery efforts in Fukushima Prefecture have hinged on the development of numerous electrical power-generation projects. For example, in the 1930s, public energy-generation projects were pursued to respond to depression, a tsunami, and frost damage to crops that devastated the region during this tumultuous period. It is also the case that Fukushima Prefecture has a long history of supplying energy to the Tokyo metropolitan region. In 1914, the Inawashiro hydropower company established long-distance transmission lines to Tokyo, and even today TEPCO continues to provide hydropower from the Inawashiro Lake region to the Tokyo metropolis. Later, in the 1970s, the Fukushima Daiichi, Fukushima Daini, and Hirono Power Plants were constructed in the prefecture, again to provide energy to the Tokyo region. Although these sites are partly responsible for stimulating local economies and employment, the development of each was advanced by operators from outside the prefecture and, especially in the case of the nuclear power plants, created an economic structure dependent on nuclear power. The largest reason for the appearance of these distorted economies is that the people of Fukushima Prefecture were not responsible for decision-making (such as halting or discontinuing operations) and lacked access to information. For example, Fukushima Prefecture had no legal basis for conducting safety inspections at nuclear power plants in the prefecture, and such inspections were only conducted under an informal agreement with TEPCO. Accordingly, it is imperative for the people of Fukushima Prefecture to become the agents of renewable energy and to have rights and responsibilities regarding these operations. With companies from outside the prefecture currently operating 80 per cent of the solar power-generation projects in Fukushima, it is clear that the people of Fukushima are not currently the leaders of these projects.

The goal of local production for local consumption targeted by Fukushima Prefecture is a major challenge that can only be pursued over the long term. While there is a need to increase renewable-energy projects over the long term, at present almost all the renewable-energy projects that have been implanted are reliant on FIT policies. Under FIT, the fixed purchase price is set to be reduced. Moreover, July 2012 to June 2015 has been designated as a premium period in which the fixed purchase price was set high. If the fixed purchase price is lowered, then incentives for implementing renewable-energy projects will be reduced and the number of new renewable-energy projects will be curtailed.[7] In order to increase renewable-energy projects over the long term, there is a need to increase the number of projects that do not rely on FIT policies, such as, for example, self-consumption. In addition, the fixed purchase price set under FIT policies will only last 20 years, at the longest. The renewable-energy projects established during the premium period will be removed from FIT in June 2035, and for existing renewable-energy facilities established in Fukushima Prefecture, the fixed purchase price will expire prior to 2040. Once the fixed purchasing period has ended, basically, operators must update their facilities and apply again for certification or remove them. Even if they want to continue operating, if the fixed purchase price is too low to justify operation, then the decision will have to be to remove them. If that happens, then the number of renewable-energy projects will decrease. Accordingly, prior to 2040, there is a fear that renewable-energy operations will rapidly decrease. For all of the above reasons, to achieve local production for local consumption of energy it is imperative to advance efforts from a long-term perspective and to think of methods that do not rely solely on FIT policies.

Acknowledgements

This study was supported by JSPS KAKENHI (Multi-year Fund) Grant Number 25220403 (Grant-in-Aid for Scientific Research (S), 2013–2017, Principal investigator: Mitsuo Yamakawa).

Notes

1 It is important to point out, however, that FIT is not the only renewable-energy policy being implemented today. A number of countries (e.g. the United States, Australia, and also Japan until June 2012) have introduced a RPS (Renewable Portfolio Standards) system that requires energy companies to use a fixed amount of renewable energy.

2 Formally, the *Act on Special Measures Concerning Procurement of Electricity from Renewable Energy Sources by Electricity Utilities*.

3 Based on an interview at the German Cooperative and Raiffeisen Confederation (Deutscher Genossenschafts- und Raiffeisenverband e.V.) in February, 2014.

4 In contrast, the stagnation of solar-power growth in Japan prior to 2011 is attributed to the government's emphasis on nuclear power and the termination of subsidies for solar power (Chowdhury *et al.* 2014).

5 Organization website: www.fre-net.jp/

6 However, it is also basically the case that there are areas in which renewable energy-generation projects cannot be developed due to restrictions in the Agricultural Land and the Urban Planning Law, and renewable energy cannot be developed here even by the landowners.

7 The Fukushima prefectural government has set 2040 as its target for realizing an energy system based on the principle of *chisan-chisho*. As of the end of March 2013 (end of fiscal year 2012 in Japan), the total cumulative amount of renewable energy installed under FIT in Fukushima Prefecture was 401,698kW (Fukushima Prefecture 2013). This amount is 99.9 per cent of the forecast amount of installation for the fiscal year of 2012, suggesting that introduction is advancing as planned. At the same time, this amount fulfills only 22.1 per cent of the goal of providing 100 per cent of local energy demand through locally generated renewable energy by 2040.

References

Agency for Natural Resources and Energy. 2013. *Enerugi Kosuto to Keizai-Eikyo Ni Tsuite* [On Energy Costs and Economic Impact]. Tokyo: Agency for Natural Resources and Energy.

Agency for Natural Resources and Energy. n.d. "Kotei Kakaku Kaitori Seido Joho Kokai yo Website" [Website for Public Information Disclosure on the Feed-in-Tariff Systems]. Accessed November 12, 2015. www.fit.go.jp/statistics/public_sp.html.

Alireza, Aslani, and Ali Mohaghar. 2013. "Business Structure in Renewable Energy Industry: Key Areas." *Renewable and Sustainable Energy Reviews* 27: 569–575.

Chowdhury, Sanjeeda, Ushio Sumita, Ashraful Islam, and Idriss Bedja. 2014. "Importance of Policy for Energy System Transformation: Diffusion of PV Technology in Japan and Germany." *Energy Policy* 68: 285–293.

Federal Ministry for the Environment, Nature Conservation and Nuclear Safety (BMU). 2011. "Renewable Energy Sources in Figures." Berlin: Federal Ministry for the Environment, Nature Conservation and Nuclear Safety (BMU). Accessed December 22, 2015. www.bivica.org/upload/energia-fuentes-renovables.pdf.

Federal Ministry for the Environment, Nature Conservation and Nuclear Safety (BMU). 2012. "Renewable Energy Sources in Figures." Berlin: Federal Ministry for the Environment, Nature Conservation and Nuclear Safety (BMU). Accessed December 22, 2015. www.fes-japan.org/wpcontent/uploads/2013/04/broschuere_ee_zahlen_en_bf.pdf

Federal Ministry for the Environment, Nature Conservation and Nuclear Safety (BMU). 2013. "Renewable Energy Sources in Figures." Berlin: Federal Ministry for the Environment, Nature Conservation and Nuclear Safety (BMU).

Fukushima Minyu. 2014. "Taiyoko Hatsuden Kennai Kigyo Sannyu Niwari Ika" [Less than 20% of Solar Power Generation by Fukushima-Based Companies]. *Fukushima Minyu Shimbun*, April 23, 2014. [In Japanese]

Fukushima Prefecture. 2013. *The Current State of the Introduction of Renewable Energy in Fukushima Prefecture*. Fukushima City: Fukushima Prefecture. Accessed December 21, 2015. www.pref.fukushima.lg.jp/download/1/h24dounyuujisseki.pdf.

Fukushima Prefecture. n.d. *Situation of the Tourism Industry*. Accessed November 12, 2015. www.pref.fukushima.lg.jp.e.od.hp.transer.com/site/portal/list277-876.html.

International Energy Agency (IEA). 2008. *Deploying Renewables, Principles for Effective Policies*. Paris: OECD.

Japan Photovoltaic Energy Association, Reconstruction Center (JPReC). Various years. *Saisei Kano Enerugi Hatsuden Setsubi tou Donyu Sokushin Shien Fukko Taisaku Jigyo Hojokin: Heisei 23 Nendo—Heisei 27 Nedo Saitaku Jiseeki Deta* [Subsidies for the Introduction of Renewable Energy Power Generation Facilities: Data on Project Approvals for 2011–2015 fiscal years]. Accessed December 21, 2015. www.jpea.gr.jp/jprec/. [In Japanese]

Kingston, Jeff. 2013. *Critical Issues in Contemporary Japan*. London: Routledge.

Menanteau, Philippe, Dominique Finon, and Marie-Laure Lamy. 2003. "Prices Versus Quantities: Choosing Policies for Promoting the Development of Renewable Energy." *Energy Policy* 31(8): 799–812.

National Institute for Environmental Studies, Center for Material Cycles and Waste Management Research. 2014. "Hosyasei Busshitsu Osen Haikibutsu no Shokyaku Shori ni kansuru Kagaku teki Chiken" [Scientific Knowledge about the Incineration of Waste Contaminated by Radioactive Materials]. Accessed April 28, 2014. www.env.go.jp/jishin/rmp/conf/waste_safety01/mat07.pdf.

Ohira, Yoshio. 2013. "Chiiki Saisei ni Muketa Fukushima-ken no Saiseikano Enerugi Seisaku ni kansuru kosatsu" [A Study About Renewable Energy Policy in Fukushima for Regional Revival]. *Journal of Public Utility Economics* 65(2): 29–36. [In Japanese]

Oshima, Kenichi. 2010. *Saiseikano Energy no Seiji Keizaigaku* [Political Economics of Renewable Energy]. Tokyo: Toyo Keizai. [In Japanese]

Oshima, Kenichi. 2013. *Gempatsu ha Yappari Wari ni Awanai* [Nuclear Power is Not Worth it After All]. Tokyo: Toyo Keizai. [In Japanese]

Samuels, Richard J. 2013. *3.11*. Ithaca, NY: Cornell University Press.

11 Beyond developmental reconstruction in post-Fukushima Japan

Daisaku Yamamoto and Mitsuo Yamakawa

Shattered windows and broken roofs still dot the landscape of Tomioka Town, a coastal community that was divided into three evacuation zones after the nuclear accident; some of the entry restrictions on the town were due to be lifted in 2016 (Figure 11.1). This edited volume has focused on the process of *fukko* (recovery, reconstruction, and redevelopment) in Hama-Dori, the region of Fukushima most severely afflicted by meltdown and hydrogen explosions at the TEPCO Fukushima Daiichi Nuclear Power Plant. Five years after the accident, one still finds a stark contrast between the *fukko* process in tsunami-hit areas of Tohoku and the areas of Fukushima affected by nuclear fallout. The ongoing nuclear disaster continues to bring dismal news, including, most notably, the discharge of radioactively contaminated water from the plant. With no end to the disaster in sight, it seems likely that it will be nearly impossible for Tomioka Town residents to happily return to their homes even when restrictions are lifted. All of the chapters in this volume clearly illustrate that while efforts to restore the livelihoods of residents in nuclear disaster-afflicted areas began soon after the accident, the hard reality is that the progress of reconstruction and redevelopment has been hampered and complicated by the continuously changing conditions of the disaster (Yamakawa and Yamamoto 2016).

Figure 11.1 Scenes of Tomioka Town in June 2016.

Note: Some of the tsunami-hit sections of the town remain little changed from March 11, 2011 (left). Some parts of the town remain off-limits, and only former residents can enter these areas during the day (right). Photos by Daisaku Yamamoto.

As we have pointed out in the concurrently published *Unravelling the Fukushima Disaster*, the disaster not only inflicted far-reaching physical, physiological, and economic damage, but also severely eroded public confidence in expert knowledge and the kind of science on which the "nuclear village" and its power had been based. Accordingly, we insisted on continuing to ask ourselves what constitutes truly useful knowledge for supporting the reconstruction and redevelopment process. We sought to identify and generate such knowledge by closely observing unfolding events and responses on the ground. Even though this is still work in progress, we believe that we can now reflect not only on the nature of the accident, but also on the process of *fukko* during the first several years. Furthermore, we can now contemplate the broader implications of reconstruction, including how the nuclear accident—and ways in which the state, TEPCO, and other major actors responded to it—is affecting other nuclear power plants and host communities in the country. In this chapter, by drawing on the chapters of this volume and other evidence, we demonstrate that policies and actions consistent with the logic of developmentalism have been selectively adopted in the post-Fukushima reconstruction process. By doing so, we also hope to point to better futures for Fukushima and for other localities hosting nuclear facilities.

The developmental regulatory system and post-disaster reconstruction

The developmental regulatory regime

The dominant institutional structure of Japan and some other Asian states during their rapid economic growth since the end of World War II has been characterized as a developmental state or developmental regulatory regime (e.g. Johnson 1982; Gao 1997; Woo-Cumings 1999; Deans 1999). In recent economic geographic literature, neoliberalization has become an encompassing concept that describes the general tendencies of recent capitalist institutional transformation in various national contexts (e.g. Harvey 2005; Brenner *et al.* 2010). However, for the mentioned (primarily east and southeast) Asian countries, including Japan, such transformations are taking place on the path-dependent basis of developmental regulatory regimes, rather than Atlantic Fordist–Keynesian regimes (cf. Kim 1999; Park 2008; Tsukamoto 2011, 2012; Park *et al.* 2012). Indeed, Fujita and Hill (2012) characterize Japan's current regulatory system as post-developmentalism, rather than neoliberalism. The notion of developmentalism therefore can arguably continue to offer a useful lens through which we understand the reconstruction process after the Great East Japan Earthquake Disaster.

In this chapter, following Yasusuke Murakaki, developmentalism is defined as

> an economic system that takes a system of private property rights and a market economy (in other words, capitalism) as its basic framework, but that makes its main objective the achievement of industrialization (or a continuous

> growth in per capita product), and insofar as it is useful in achieving this objective, approves government intervention in the market from a long-term perspective.
>
> (Murakaki 1999, 145)

Developmental policy interventions take the form of both growth policies and redistributive policies, although the latter are often overlooked in the discussion of East Asian developmentalism (Suehiro 1998). One of the distinct aspects of developmental redistributive policies is that they are often linked to productive activity (Murakami 1999; Estévez-Abe 2008). That is, developmental redistributive policies tend to redistribute income to economic actors (e.g. business owners, firms, and local governments) in order to preserve and promote productive activities, as opposed to the provision of "lump-sum" payments, a feature of an ideal form of welfare state, to bolster the consumptive power of individuals in need. Japanese regional developmental policies have long taken the form of mini-growth projects and interventions such as the construction of productive infrastructure in, and channeling private-sector investments to, rural peripheries (cf. Kawashima and Kamozawa 1988; Calder 1988; Murakami 1999).

For our discussion, the literature on developmentalism helps to formulate several expectations about the course of post-disaster reconstruction in Japan. Since faltering national economic growth is the most serious source of legitimation crises for a developmental state (Deans 1999), problems in need of policy solutions (i.e. disaster damage that needs to be solved) will likely be framed as a lack of economic growth, and the policy focus will be on making afflicted regions and sectors productive once again rather than focusing on compensation for and the welfare of individuals. This chapter attempts to substantiate this claim by highlighting the government's emphasis on decontamination and the return of evacuees, industrial projects in afflicted regions, and massive and coordinated capital investment to enhance safety measures for existing nuclear power plants while leaving emergency evacuation plans behind. More importantly, however, our goal is to show, when this form of developmental reconstruction proves to be in any way problematic, how we may guide the course of reconstruction in Fukushima and its effects on other regions in what we view as a more desirable direction.

Reconstruction in Fukushima

The present reconstruction scheme for nuclear disaster-afflicted regions is first and foremost characterized by intensive decontamination work (Figure 11.2), while the compensation and support provided to disaster victims and evacuees does not adequately respond to their real needs and concerns (Chapters 1 and 3). We do concur that for those evacuees who have already returned or plan to return to their homes in radioactively contaminated areas, lowering the radioactivity rate in their communities is absolutely essential (Yamakawa and Yamamoto 2016; Ogawa 2013), even though the effectiveness of such operations can be questionable in some cases (e.g. Edgington 2016). What is clear, nevertheless, is that state-supported

Figure 11.2 Landscapes of decontamination bags.

Note: Among the most symbolic landscapes of the present reconstruction scheme for nuclear disaster-afflicted regions are the hundreds of thousands of large bags containing radiation-contaminated soil and organic materials piled up on fields and in valleys of the afflicted areas. In the photograph on the right, the bags are covered by large green sheets. Photos taken in Iitate Village in 2015 by Daisaku Yamamoto.

decontamination projects essentially preserve the structure of interregional redistributive mechanisms based on public civil-engineering projects that has long characterized the Japanese developmental regulatory regime (Fujimoto 2016). In this scheme, general construction companies headquartered in Tokyo are at the top of the hierarchy, while the actual work is contracted out to second and third-tier subcontractors who often hire day laborers. For Fujimoto (2016), decontamination projects are "even worse" than conventional public infrastructure projects because they have limited economy-wide productivity-enhancing effects. Furthermore, perhaps most problematic is that, as Sato (Chapter 3) demonstrates, residents' concerns and voices—including questions concerning the effectiveness of decontamination work and requests to review reconstruction budgets that heavily favor decontamination and return over support for relocation—remain unaddressed and carefully excluded from policy discussion. In other words, there is little evidence that decontamination-intensive reconstruction is grounded in solid societal consensus.

There is no question that the revitalization of economic activities and the creation of employment opportunities are critical concerns, and on the surface the policies and projects introduced over the past five years appear to directly address these concerns. Fukushima Prefecture established the "Vision for Revitalization in Fukushima Prefecture" in August 2011, five months after the disaster. The Vision proposed three overarching goals: 1) building a safe, secure, and sustainable society free from nuclear power; 2) revitalization that brings together everyone who loves and cares about Fukushima; and 3) a homeland we can all be proud of once again. Based on the Vision and the national government's Act on Special Measures for the Reconstruction and Revitalization of Fukushima, the prefecture outlined "Industrial Revitalization Plans" in May 2013. The plans' action programs identify key sectors—including the renewable-energy industry (cf. Chapter 10), the medical industry, and the information and communication technology (ICT) industry—and activities, including the establishment of an environmental radiation-monitoring and research institute (euphoniously called the "Environmental Creation Center") and the Coastal Region Agricultural Revitalization Research Center. The targets of these programs may appear reasonable, if not quite ambitious.

However, a closer examination of the proposed projects and related policy documents reveals problematic issues. In terms of renewable-energy industry development, the Fukushima Renewable Energy Institute was opened in Koriyama by the National Institute of Advanced Industrial Science and Technology. The Fukushima Floating Offshore Wind Farm Demonstration Project has been carried out by a consortium of major corporations. These projects may seem indicative of a renewed and strong emphasis on the development of renewable energy, and may even be interpreted as a suggestive move away from nuclear power.

However, upon reading the Fourth Strategic Energy Plan (April 2014) issued by the Ministry of Economy, Trade and Industry (METI), which supposedly reflects the lessons learned from the nuclear disaster, one would see a very different picture. The previous third plan, drawn up in June 2010, had emphasized "stable energy provision" (energy security), "cost considerations" (economic efficiency), and "environmental considerations" (climate change countermeasures). Following the Great East Earthquake Disaster, the plan was amended to emphasize "safety considerations." However, the newly crafted fourth plan adds "international energy cooperation" and "economic growth" as additional considerations. The plan asserts, for example:

> In the field of nuclear power, the Japan-U.S. Bilateral Commission on Civil Nuclear Cooperation was launched in order to further enhance cooperative relations between Japan and the U.S. after the TEPCO Fukushima nuclear accident. Concerning the system to support use of nuclear power, Japanese and U.S. reactor manufacturers have already established a framework for expanding business in an integrated manner in the commercial field as well. Japan and the U.S., as partners, play a significant role in enhancing a global system for nuclear use while internationally ensuring peaceful use of nuclear power, nuclear non-proliferation, nuclear security, and so on.
>
> (METI 2014, 79)

Furthermore, the plan emphasizes the importance of taking advantage of "low cost energy supply by enhancing economic efficiency … [as] a precondition for keeping existing business operations in Japan and attaining further economic growth" (18). It further calls for

> an opportunity for Japan's energy industry to strengthen its competitiveness and boost its presence in the global market. It is expected to contribute to improving the trade balance through exports by energy-related companies of energy-related equipment and services with high value added.
>
> (18)

These statements by national-level policymakers therefore pave the way for continuous reliance on nuclear energy.

Additionally, in terms of strategic support for the medical industry, we are concerned that medical *industrial* promotion has become an end in itself, rather than being a means to protect and foster the physical and mental health of those who were affected by the disaster. In its effort to promote the medical-industry cluster, the Fukushima Medical Device Development Support Center (which will be the first center of this kind in Japan), where comprehensive safety assessments of medical devices can be undertaken (Medical Industry Cluster 2016), was established with support from the central government. In addition, the Fukushima Global Medical Science Center was established at Fukushima Medical University in order to "continue the Fukushima Health Management Survey, establish cutting-edge medical facilities and treatment modalities, foster globally-oriented medical professionals, and promote the medical industry to rebuild and revitalize our community" (Fukushima Medical University 2016, 1). Nevertheless, seeing the declining screening rate of the Health Management Survey year by year (Yamakawa and Yamamoto 2016), we cannot help but feel that actual care for the disaster victims' health may be becoming a secondary focus.

Similar problems are embodied in other projects, including the ICT-enabled "smart city" project in Aizu-Wakamatsu City, the "Environmental Creation Center" in Miharu Town, and the Coastal Region Agricultural Revitalization Research Center in Minami-Soma City. These projects all share common characteristics: the main actors of these projects are companies and institutions headquartered in Tokyo, and proponents of these projects only wishfully conjecture that these centers will stimulate local industries. Yet, it is far from clear how these projects would actually facilitate the growth of those small and medium-sized manufacturing firms in Minami-Soma (Chapter 8), and how they may contribute to the rebuilding of small commercial businesses in Kawauchi (Chapter 9). The current rhetoric strongly evokes the logic employed when nuclear power plants were built in rural peripheries of the country.[1]

Impact on other nuclear-host regions

Not surprisingly, the nuclear accident at the Fukushima Daiichi Nuclear Power Plant has had significant effects on other nuclear power plants in the country, and more broadly on the communities in which these plants are located. Currently, most nuclear power plants are under maintenance, as the new regulatory requirements (including revised safety standards) enacted on July 8, 2013 require that all nuclear power plants must meet enhanced standards before restarting commercial operation. Accordingly, the nine commercial-power companies with nuclear power plants are currently retrofitting their plants in order to withstand large-scale natural disasters (e.g. tsunami, tornadoes, and volcanic eruptions), terrorist attacks, and severe accidents that had not been considered under old regulations. If one visits any of the nuclear plants in Japan today, one will see massive seawalls being built around plants, the addition of emergency power sources and fire engines, and clear-cut of trees around plants. TEPCO, for example, appears eager to restart its Kashiwazaki-Kariwa plant in Niigata Prefecture, which, with seven reactors, has the largest power-generation capacity in the world at a single nuclear power plant; the company is running a television-advertisement campaign (targeted only at viewers in Niigata) to communicate its efforts to improve safety measures at the plant. In fact, some local media outlets report that local construction companies and some of the affiliated companies of TEPCO are unusually busy, leading to an increase in local employment and transactions in nuclear-host localities, which is a somewhat ironic situation. What the current development shows is that there are few institutional barriers against, and strong state support for, these capital investments by the private sector.

On the other hand, progress on designing effective evacuation plans in the case of nuclear emergency, another pillar of countermeasures for nuclear accidents, has been slow. Under the Disaster Management Basic Plan, revised in September 2012, local municipalities containing areas within 5 km (Precautionary Action Zone: PAZ) or 30 km (Urgent Protective Action Planning Zone: UPZ) of a nuclear power plant are required to establish evacuation plans. Under the current law, the responsibility to create evacuation plans is delegated to local municipal governments. The national government provides "support" to those municipalities that are developing their own plans (Cabinet Office 2016). To some local-government officials, the current stance of the national government appears to indicate evasion of responsibility and a lack of commitment. For example, Kashiwazaki City formulated its "Wide-Area Evacuation Plan in Case of Nuclear Emergency" but states on its website that "the evacuation plans still have many problems. We wish to make the plans more effective by listening to your opinions, and we ask for your cooperation to that end." (authors' translation; Kashiwazaki 2015). Some residents are concerned that the plant may restart its operation before the "more effective" plans are implemented.

In summary, the current post-nuclear disaster reconstruction scheme is characterized by prioritizing decontamination projects over thorough support of victims and evacuees, exognenous industrial growth over more locally oriented development, and capital investment in plant safety measures over nuclear-emergency evacuation planning (cf. Chapter 2). What we can see here is a reconstruction scheme that perceives the lack of productive capacity in focal regions as the most important problem to be addressed, and that aims to solve "the problem" by facilitating economic growth and productive activities (in which external actors often play the central role) without necessarily supporting and empowering truly disadvantaged individuals and social groups. The embodied logic is very much consistent with that of the developmental regulatory regime.

Accordingly, it is important to keep in mind that a move away from nuclear power alone does not automatically alter the structure of dependence that is embedded in the developmental regulatory regime. For example, an increasing number of nuclear reactors are reaching the end of their designated lifespan of 40 years, and power companies have announced that some commercial reactors will be decommissioned. It is unclear at this moment whether the pace of decommissioning will accelerate, but local nuclear-host governments are now more keenly aware of the prospects of losing various forms of nuclear-related grants and subsidies. On this point, the Group of Zero Nuclear Power, a bipartisan group of 64 legislators, argues for the need for financial compensation for nuclear-host localities in order to promote nuclear decommissioning (Group of Zero Nuclear Power 2013). However, such a solution alone risks reproducing the very structure of dominance and dependence that promoted nuclear power as a regional-development policy under the developmental regulatory regime.

Beyond the developmental regulatory regime

One of the most striking differences between Japan and the United States, two countries that are integrally related in terms of nuclear development, in the development of civilian nuclear power plants is the way in which Japan's nuclear power plants are tied rhetorically and substantively to the idea of regional economic development. Nuclear power plants have been advertised as promoting local economic growth through job-creation and multiplier effects. The establishment of the Power Source Siting Laws (1973) instituted the provision of fiscal subsidies to local governments, which often resulted in the building of a disproportionately large number and expanse of public facilities compared to other similar-sized localities. Besides nuclear power plants, many rural peripheries in Japan were swept by "development projects" during the postwar period, as seen in the construction of massive dams, heavy-industry complexes, high-tech industrial parks, and leisure resorts. These projects are fundamentally for the urban core (i.e. provision of electricity, inexpensive land for branch plants, and holiday destinations for urban workers), but are promoted as benefitting the rural peripheries (this is staunchly their secondary purpose). Some of these projects did bring in external investment and created jobs in rural peripheries at least for a time, but many ultimately did not meet their original goals and aspirations, as evidenced, for example, by persistent and accelerating depopulation in an area where a dam was built nearby (Machimura 2006); environmental pollution (Broadbent 1998); vacant plots of land for factories, empty resort hotels, and condominiums (McCormack 2001); and concerns over the lack of synergy between power plants and local manufacturers (Okada and Kawase 2013).

What are being sought today, nevertheless, are effective ways to move beyond the developmental regulatory regime, rather than simply reflectively criticizing what has already happened. In particuar, we must tackle at least two interrelated challenges. The first is quite obvious: the system subjects local economies to dependence on external money and control, and over time local residents' livelihoods become integrated in a complex manner with externally controlled establishments and what comes with them (e.g. subsidies and grants). Therefore, in order to carve out more self-reliant and sustainable local development trajectories, it is essential to cultivate local economies that do not depend excessively on non-local, external investments and controls.

A second point, and one which may not be as obvious, is that developmental projects (e.g. nuclear power plants and golf resorts) tend to be socially divisive because the process of negotiation and decision-making tends to generate corruption, suspicions of nepotism, and antagonism between different strata of local residents (e.g. landholders and non-landholders) (Yamamoto and Yamamoto 2013). These social divisions can be long-lasting (even when such projects are not, in the end, realized). It is therefore essential to cultivate local capacity to foster critical engagements among residents and to reach informed consensus.

Building alternative economies

One of the emergent discourses of regional development is strengthening of the intra-regional circulation of income, investment, and resources. Proponents of this idea argue that conventional regional-development schemes typically rely on extra-local investments and markets, and end up with a leakage of profits back to urban cores (e.g. Okada 2005; Fujiyama 2015). Fujiyama (2015), for example, argues that a regional economy can grow by replacing small fractions of "imported" goods and services (e.g. bakery goods from outside of the region) consumed in a focal region with locally produced equivalents over time. This is obviously called import substitution, and has lost its popularity as a dominant development discourse internationally since the 1980s. The idea has, nevertheless, been revisited by scholars such as Markusen (2007), who traces the importance of the notion to the works of Tiebout (1956). While this kind of endogenous development theory is certainly attractive, what interests us is how these ideas actually play out on the ground, and what lessons we can learn from them.

Let us take a look at one local effort to commerce renewable-energy business in Fukushima (Yamakawa 2016). Tsuchiyu Onsen is an administrative area within Fukushima City, located on the western side of the city, and is known for its hot springs and inns. After the Great East Japan Earthquake Disaster, 2 of the 16 inns permanently closed their doors, and one suspended its operations indefinitely. For a while the town was busy, because it accommodated 900 nuclear-accident evacuees, but as they moved on to temporary housing elsewhere the number of guests and visitors declined, and three more inns closed. The population of the town dropped from 465 to 435 after the disaster. The town proposed a "Tsuchiyu Onsen Smart Community Draft Plan" in May 2012, in order to respond to the struggling local economic situation. It set the development of renewable energy as one of the main components of the plan, and aimed to implement two projects: a binary cycle plant, a type of geothermal power plant using hot spring water and steam in the area, and a small hydroelectric power plant, using the already available erosion-control dam bank. According to the plan, the plants would produce enough electricity for the town (about 1,000 kW), and might also be able to sell excess power to outside areas. The project plans were originally drafted by a group of 29 local individuals in October 2011, and a locally funded corporation (*Genki Appu Tsuchiyu* Co.) was established to run the business in October 2012. The company now employs four people, three of whom are young college graduates from outside of Tsuchiyu Onsen. The two power plants began operating in May and November 2015. While it is still too early to tell whether these projects will be successful and contribute to the development of a more self-reliant and sustainable local economy, we can identify a few important lessons.

First, the introduction of renewable energy could be an opportunity to develop more self-reliant local economies, but this is not guaranteed. The most important point to note is that this business project was initiated and carried out by local stakeholders. Previous studies indicate that similar renewable-energy business brings significantly different levels of profits to localities depending on how the business is carried out (Funabashi 2012; Yamakawa 2016). For example, a non-local firm built ten wind turbines (total output of 13,000 kw) in a village in Tokushima Prefecture and recorded sales of 267 million yen (2005), but the net revenues to the village were approximately 7.5 million yen after a reduction in the national government's local grant tax that accounted for the plant's property-tax revenues (about 30 million yen). On the other hand, a locally owned and operated plant with two wind turbines (1,200 kw) in a town in Kochi Prefecture resulted in 23.2 million yen of revenue for the town. In the case of Tsuchiyu Onsen, too, it is critical that the agency that drove the project did not "come from above." Indeed, multiple chapters in this volume show the degree to which taking advantage of the locally based knowledge, technologies, and organizational skills and experiences of local actors, cultivated and nurtured prior to the disaster, has been critical in the process of rebuilding livelihoods and sustaining community ties during crisis (Chapter 4); recovering transportation services more effectively (Chapter 5); keeping food systems more resilient against stigmatization (Chapter 7); and implementing renewable-energy projects (Chapter 10).

Second, having said that, we would also point out the practical challenges to carrying out projects like this. Capital-intensive projects like the establishment of power stations cannot usually be paid for by local investors alone. The estimated cost of facility construction for the Tsuchiyu Onsen project was 630 million yen, and 570 million yen of the total cost was secured through a grant from government-affiliated agencies. An insistence on exclusively locally funded ventures would have meant there would have been no way to move forward with the Tsuchiyu Onsen project. To us, this episode shows the importance of approaching post-nuclear and post-developmentalist local economic development from the perspective of the "weak theory of economy" (Gibson-Graham 2006). That is, we must refuse to see cooptation, such as accepting funding from governments and institutions that may not share the same values and goals, as a necessary condition of consorting with power. Instead, we accept it as "an ever-present danger that calls forth vigilant exercises of self-scrutiny and self-cultivation—ethical practices, one might say, of 'not being co-opted'" (Gibson-Graham 2006, xxvi). Therefore, what matters most is not simply the source of funding or resources, but where the main agency is and whether it can remain reflexive. The importance of such pragmatic uses of non-local funding, investment, and resources is demonstrated in other chapters of this volume (e.g. Chapters 8 and 9).

Third, we must accept that socio-cultural factors and dynamics may matter hugely (in some cases more than economic logic) in carrying forward local projects like this. For the Tsuchiyu Onsen project group, one of the remaining questions was how to secure the remaining 60 million yen—the difference between the total cost (630 million yen) and the grant received (570 million yen). At first, an executive director of a major regional bank in Fukushima came to work on the project, but his approach was perceived by the Tsuchiyu locals as too top-down and heavily reliant on large, nonlocal companies, and his attitude was seen as too haughty. In the end, a manager of the finance division of a smaller local bank, who happened to be from Tsuchiyu Onsen, was able to gain the trust of the locals, and played a pivotal role in securing the 60 million yen from his bank for the project. It is hard to know in retrospect whether this was the best economic decision (e.g. which was a better lending scheme), but this kind of local sentiment is real and must be respected. As shown in other chapters of this volume, we have seen many instances in which trust among key actors, which must be cultivated over time, played a decisive role in rebuilding and sustaining social and economic organizations (e.g. Chapters 4 and 7).

Our final point is not directly related to the Tsuchiyu Onsen project, but is instead more general. The discourse of developmentalism typically reduces the meaning of local "development" to easily quantifiable measurements. "Development" is justified by the number of jobs it would create, and by sales and income generated. Yet it is this very simplification of the notion of "development" that is at the heart of the problem. As Noda (Chapter 4) shows, the stream water for a small hamlet is not simply a commodity that satisfies the physiological requirements of residents (and hence is substitutable with well water or city water); rather, it has the social value of maintaining community bonds, and thus is not easily substitutable (also see Kaneko 2016). It is therefore critical to evaluate the Tsuchiyu Onsen project not only in terms of the number of jobs or revenue created, but also by paying close attention to how the project reshapes the social dimensions of the community, including how those young, new residents working for the renewable-energy project may have impacts on a broader range of community development (*machi-zukuri*) issues.

Building foundations for deeper local social engagement

In July 2012, as part of drafting the Energy Strategic Basic Plans, there was a national-scale deliberative poll to gauge public attitudes about different future energy options (Sone *et al.* 2013; Executive Committee of the Deliberative Poll on Energy and Environmental Policy Options 2012a, b). This was one of the first large-scale deliberative polls in Japan, a practice informed by the idea of deliberative democracy (Fishkin 2009). Three scenarios put forward included the "zero scenario" (reducing nuclear power to 0 per cent of energy supply by 2030), "15 scenario" (reducing nuclear to 15 per cent by 2030), and "20–25 scenario" (reducing and keeping nuclear at around 20–25 per cent by 2030). In August 2013 there was a debate forum; as the forum progressed, participants' support for the zero scenario increased from 37 per cent to 47 per cent, while support for the 20–25 per cent scenario was unchanged at 13 per cent. In addition, the per centage of those who said they "support multiple scenarios" or felt "no strong support for any of the scenarios" declined from 38 per cent to 25 per cent. This deliberative poll framed the problem of energy in terms of safety concerns, energy security, effects on global climate change, and costs, and failed to address some critical ethical issues such as who bears the costs/risk of (nuclear) energy and (radioactive) waste. Nevertheless, this social experiment did show that more informed (trans) formation of citizens' opinions is possible through carefully designed opportunities to learn and debate.

The Citizens' Commission on Nuclear Energy (CCNE 2013) indeed argues for such national forums, in which a variety of opinions are brought out and debated to form a more genuine national consensus as a basis for policies on energy. Although their argument is well taken, institutionalizing such national forums alone would be insufficient in shaping energy options, especially when it comes to the issues involved with nuclear energy. This is because, in addition to domestic political and administrative organizations, the US–Japan Alliance and nuclear-host localities hold critical power over nuclear-energy politics in Japan. Unlike some other countries, such as the United States, nuclear-host localities hold de facto veto power over the construction, modification, and even abandonment of nuclear facilities (CCNE 2013, 14–15).

In the media, nuclear-host localities are often portrayed as "addicted to nuclear money" and as selfishly demanding the reactivation of reactors, most of which, as noted, are currently idle for the purposes of testing under stricter safety regulations. Yet, whether such images truly reflect a genuine local consensus is far from clear. We must first understand the nature of decision-making and consensus-building at local scales, because the apparent "resistance" of nuclear-host localities to national policies may not in fact reflect the consensus of the local residents. We must also question whether the kind of national forums that the CCNE endorses can realistically function at local levels in which non-anonymous, dense human interactions often prevail.

To illustrate these points, let us draw on the study by Yamamuro (1998), who examined the controversies over a planned nuclear plant in Maki Town, Niigata Prefecture, during the 1990s. During the controversy, a local citizen group organized a volunteer-based resident vote over the planned project.[2] One of the women that Yamamuro interviewed had a daughter who used to work at the plant, as well as other relatives who were working for the power company (Tohoku Electric Power Company) that proposed the nuclear plant project. She said:

> I have been told not to go voting and I pretended that I was in favor of the project. So, I would have been in trouble if someone saw me going to the voting. So, I carefully chose the timing of voting, early in the morning, so that few would see me there. But, if I had been found, I was going to say, "anyone can go vote; it does not matter whether you are for it or against it, and I voted 'yes' to the project)." You know, I'm in the position to think about (my daughter) being the wife in a household and various other restraints.
>
> (Yamamuro 1998, 196; authors' translation)

From this quote, one can see the magnitude of "restraints" (which are essentially norms for carrying out ordinary daily lives) and how power, rather than being some abstract entity, is actually exercised on the lives of people in concrete forms. One can also appreciate potential obstacles to that kind of deliberative democracy, advocated as an ideal form of public consensus-making. Below we discuss three issues regarding what must and can be done to "hear" more genuine voices of residents in nuclear-host localities.

First, we must take advantage of existing institutions for critical local engagement. Among the worst-case scenarios for nuclear-host localities would be one in which the national government carelessly imposes a local version of the deliberative polls for future energy options, disregarding formal and informal instutitions that have been developed in localities over time. It is first and foremost critical to carefully examine what kinds of institutions are already there to "hear" the minds of local residents. For example, in Kashiwazaki City, there is an organization called the "Local Council to Secure the Transparency of the Kashiwazaki-Kariwa Nuclear Power Plant" (hereafter the "Local Council"), which was established after the TEPCO corporate scandal in which the company falsified repair records at its plants in Fukushima and Kashiwazaki-Kariwa in the early 2000s. The Local Council, after a rocky start in 2002, has become a rare place in which representative members from local advocate and opponent groups and from various local organizations that do not take a clear position, as well as representatives from TEPCO and local governments (Niigata Prefecture, Kashiwazaki City, and Kariwa Village), meet on a monthly basis with regard to various themes related to the plant. The Council explicitly states that it does not aim to reach consensus over nuclear power or the plant itself, beyond shared agreement over the pursuit of transparency to secure safety.

Nearly 13 years since its inception, the Local Council now faces many problems and concerns, such as a lack of conversation among Council members and the risk of becoming overly formalized and forgetting its original motivation. Nevertheless, the fact that the Council has met more than 150 times to date provides it with a certain level of legitimacy both internally and externally. One of the Council's members, representing a pro-nuclear organization, says:

> I'm in the position to support nuclear power. But because of my position, when there is a problem in TEPCO [implying the most recent scandal in which the company came to admit, after five years of the accident, that there was actually an operational manual, which would have allowed it to declare a meltdown during the onset of the accident (*Asahi Shimbun*, 2016)], my critical comments at a regular Council meeting become more effective [than those of opponent groups], I think.
>
> (Interview with a Local Council member, May 12, 2016)

In these ways, the Local Council has become a place in which the residents' more nuanced thoughts are revealed. It is therefore vital to take advantage of local institutions such as the Local Council, to seek concrete ways to deepen the level of critical engagement within, and to build new institutions and organizations from existing ones, rather than replacing them.

Second, small innovations and "tricks" matter. There have been hundreds and thousands of various meetings locally, regionally, and nationally that pertain to the reconstruction and redevelopment of nuclear disaster-afflicted areas. In many of these meetings and forums, there are potential opportunties to "hear" genuine voices of the members, but some of these opportunities are never realized (e.g. Chapter 3). One of the authors (Yamakawa) has served on a number of committees related to post-Fukushima reconstruction over the past five years, and has learned the importance of seemingly trivial measures that are actually crucial in hearing the voice of meeting attendees. For example, in most meetings, few would volunteer to voice their opinions if the moderator asks "are there any opinions?" (or "what opinions do you have?," as teachers in North America are nowadays encouraged to do). In the Japanese context, a more effective method is to explicitly name each person and ask their opinion; they will then say, "because I am called by the chairperson/moderator, I would hesitantly express my opinion..." Moreover, this is particularly effective during the first meeting (if meetings are to be held on more than one occasion), because members often express their most personal thoughts and feelings (e.g. "I am really sad that I cannot see my grandchildren who evacuated away") at that time. What often happens is that after the first meeting, the organizations represented by the members begin to influence the members' opinions, and their voice increasingly becomes that of their organization. Having their most personal thoughts and feelings revealed and shared among members upfront often facilitates more productive and meaningful conversation among them later on.[3] What this illustration points out for us is that there may be various seemingly insignificant, but in reality particularly effective, ways to improve the quality of engagement among citizens, which may lead to the formation of "Fukushima-style democracy" or other locally grounded forms of critical engagement and decision-making.

Third, it is important to realize and learn how "deeper," more genuine public consensus is skillfully watered down or removed altogether in actual policies. As mentioned in the Introduction to this book, currently there appears to be a push-back against the inclusion of the "nuclear accident" in public memory. Or, as described in Chapter 1, only one of the seven principles outlined in the guidelines for reconstruction by the Reconstruction Design Council (June 25, 2011) specifically refers to the nuclear accident, and in more specific policy documents the national government sees its "responsibility" only in revitalization and reconstruction (rather than in the accident itself). The problem is that such a push-back does not take the form of overt domination or persuasion. Instead, it may take such forms as sending personnel from the national or prefectural government to an afflicted municipality to act as vice mayor ("to accelerate reconstruction"); crafting policy documents with introductions that are very emotional in tone but where concrete policies reflect the interests of relevant state agencies more than anything else; and including "expert advisors" (such as professors and researchers) in the policymaking panel after the general direction of a project has already been formed externally. Therefore, it is imperative for all of us who care about the well-being of our community to cultivate critical eyes and minds to scrutinize these tactics.

Conclusion

The central argument of this concluding chapter has been that we must be critical of the ways in which post-nuclear disaster *fukko* is being carried out, but at the same time must begin a better form of *fukko* from where we stand now. We suggest that the idea of developmentalism still provides a useful analytical lens through which we understand and systematically critique the nature of reconstruction and redevelopment in post-Fukushima Japan. Yet, that is only one side of the story; the other side, seeking practical ways in which to steer *fukko* in a better direction, is arguably far more challenging, because we cannot isolate such discussion from the historical and geographical specificities of regions and communities. The chapters included in this volume are the records of efforts and struggles related to "starting from where we stand now," which are deeply embedded in such specificities.

Acknowledgements

We would like to thank Jay Bolthouse for helpful comments and suggestions, and Angelica Greco for her assistance in editing the chapter.

Notes

1 According to the *Overview of Electric Source Siting Policies* (2004), it is estimated that the building of a nuclear power station with an output of 1.35 million kilowatts will bring a total of 89.3 billion yen (about 1.14 billion U.S. dollars) in the form of subsidies and property taxes to a host prefecture and municipalities over a period of 20 years (ten years before and after the start of operation). Despite the large face value of pecuniary compensation, the overall benefits of large power plants for regional development have been found to be short-lived and in some cases negative (Yamakawa 1987; Shimizu 1992; Okada 2012).

2 Ultimately the nuclear power-plant project was not built.

3 These insights are reminiscent of the situation described by Gibson-Graham (2006), who conducted a focus group among different stakeholders in the Latrobe Valley, a coal-mining and power-producing region of Australia, which had been going through restructuring in the 1990s. Harry, a former supervisor at the power company who acknowledged the large-scale lay-offs as an unfortunate but rational economic decision, broke down in tears when he recalled his fellow workers who had to leave their jobs:

> [t]he long embodied antagonisms between Harry as a worker and union man and the likes of the businessman who sat next to him drained away, as an empathetic atmosphere of care was established and past ideological enemies found themselves respecting each other in their honesty.
>
> (Gibson-Graham 2006, 32)

References

Asahi Shimbun. 2016. "TEPCO Official Knew Standard for Meltdown at Fukushima." *Asahi Shimbun*, April 12. Accessed June 30, 2016. www.asahi.com/ajw/articles/AJ201604120056.html.

Brenner, Neil, Jamie Peck, and Nik Theodore. 2010. "Variegated Neoliberalization: Geographies, Modalities, Pathways." *Global Networks* 10(2): 182–222.

Broadbent, Jeffrey. 1998. *Environmental Politics in Japan: Networks of Power and Protest*. Cambridge, UK: Cambridge University Press.

Cabinet Office, Government of Japan. 2016. *Chiiki Bosai Keikaku Hinan Keikaku Sakutei Shien* [Local Disaster Prevention Plans and Support for Evacuation Plan Design]. Accessed June 30, 2016. www8.cao.go.jp/genshiryoku_bousai/keikaku/keikaku.html. [In Japanese]

Calder, Kent. E. 1988. *Crisis and Compensation: Public Policy and Political Stability in Japan, 1949–1986*. Princeton: Princeton University Press.

Citizens' Commission on Nuclear Energy (CCNE). 2013. *Genpatsu Zero Shakai eno Michi – Atarashii Koron Keisei notameno Chukan Hokoku* [Road to a Nuclear-Free Society: An Interim Report for Forming New Public Consensus]. Tokyo: CCNE. [In Japanese]

Deans, Phil. 1999. "The Capitalist Developmental State in East Asia." In *State Strategies in the Global Political Economy*, edited by Ronen Palan, Jason Abbott, and Phil Deans, pp. 78–102. London: Continuum Intl Pub Group.

Edgington, David W. 2016. "How Safe is Safe Enough? The Politics of Decontamination in Fukushima." In *Unravelling the Fukushima Disaster*, edited by Mitsuo Yamakawa and Daisaku Yamamoto, pp. 79–105. London: Routledge.

Estévez-Abe, Margarita. 2008. *Welfare and Capitalism in Postwar Japan. Cambridge Studies in Comparative Politics*. Cambridge: Cambridge University Press.

Executive Committee of the Deliberative Poll on Energy and Environmental Policy Options. 2012a. *Deliberative Poll on Energy and Environmental Policy Options*. Accessed June 30, 2016. http://cdd.stanford.edu/mm/2012/jp-energy-policy.pdf.

Executive Committee of the Deliberative Poll on Energy and Environmental Policy Options. 2012b. *Deliberative Poll on Energy and Environmental Policy Options (Summary)*. Accessed June 30, 2015. www.cas.go.jp/jp/seisaku/npu/kokumingiron/dp/120827_02.pdf. [In Japanese]

Fishkin, James S. 2009. *When the People Speak*. Oxford: Oxford University Press.

Fujimoto, Noritsugu. 2016. "Decontamination-Intensive Reconstruction Policy in Fukushima under Governmental Budget Constraint." In *Unravelling the Fukushima Disaster*, edited by Mitsuo Yamakawa and Daisaku Yamamoto, pp. 106–119. London: Routledge.

Fujita, Kuniko and Richard Hill C. 2012. "Industry Clusters and Transnational Networks: Japan's New Directions in Regional Policy." In *Locating Neoliberalism in East Asia*, edited by Bae-Gyoon Park, Richard C. Hill, and Asato Saito, pp. 27–58. West Sussex: Wiley-Blackwell.

Fujiyama, Hiroshi. 2015. *Denen Kaiki 1% Senryaku* [1% Strategy of Countryside Resurgence]. Tokyo: Nosangyoson Bunka Kyokai.

Fukushima Medical University. 2016. *Fukushima Global Medical Science Center Basic Concept*. Accessed June 30, 2016. www.fmu.ac.jp/fgmsc/wp/wpcontent/uploads/2015/12/koso_gaiyou_en.pdf. [In Japanese]

Funabashi, Harutoshi. 2012. "Chiiki ni Nezashita Saisei Kano Enerugi Notameno Seido, Seisaku, Shutai Towa" [What are the Institutions, Policies, and Actors Necessary to Promote Renewable Energy Rooted in Local Regions?] Paper presented at Forum on "Chiiki no Enerugi to Okane wo Chiiki to Chikyu ni Ikasu," Fukushima, September 26.

Gao, Bai. 1997. *Economic Ideology and Japanese Industrial Policy*. Cambridge: Cambridge University Press.

Gibson-Graham, J. K. 2006. *A Postcapitalist Politics*. Minneapolis: University of Minnesota Press.

Group of Zero Nuclear Power. 2013. *Haishi Taisho Genshiro Shuhen Chiiki Shinko Nikansuru Tokubetsu Sochihoan Kosshian (May 30, 2013)* [Preliminary Draft Legislations on the Development of Regions around Decommissioning-Target Reactors]. Accessed September 12, 2016. http://genpatsu0.cocolognifty.com/blog/files/20130530chiiki.pdf. [In Japanese]

Harvey, David. 2005. *A Brief History of Neoliberalism*. Oxford: Oxford University Press.

Johnson, Chalmers. 1982. *MITI and the Japanese Miracle*. Stanford: Stanford University Press.

Kaneko, Hiroyuki. 2016. "Radioactive Contamination of Forest Commons: Impairment of Minor Subsistence Practices as an Overlooked Obstacle to Recovery in the Evacuated Areas." In *Unravelling the Fukushima Disaster*, edited by Mitsuo Yamakawa and Daisaku Yamamoto, pp. 136–153. London: Routledge.

Kashiwazaki City. 2015. *Kashiwazaki City Wide-Area Evacuation Plan in Case of Nuclear Emergency*. Accessed June 30, 2016. www.city.kashiwazaki.lg.jp/atom/genshiryoku/taisaku/kouiki.html. [In Japanese]

Kawashima, Tetsuro, and Iwao Kamozawa. 1988. *Gendai Sekai no Chiiki Seisaku* [Regional Policies in the Contemporary World]. Tokyo: Taimeido.

Kim, Yun Tae. 1999. "Neoliberalism and the Decline of the Developmental State." *Journal of Contemporary Asia* 29(4): 441–461.

McCormack, Gavan. 2001. *The Emptiness of Japanese Affluence*. Armonk, New York: East Gate.

Machimura, Takashi (ed.). 2006. *Kaihatsu no Jikan Kaihatsu no KūKan: Sakuma Damu to Chiiki Shakai no Hanseiki* [Time of Development and Space of Development: A Half-Century of the Sakuma Dam and the Local Society]. Tokyo: Tokyo University Press.

Markusen, Ann. 2007. "A Consumption Base Theory of Development: An Application to the Rural Cultural Economy." *Agricultural and Resource Economics Review* 36(1): 9–23.

Medical Industry Cluster Promotion Unit Commerce, Industry & Labour Department Business Creation Division Fukushima Prefectural Government. 2016. *A Greeting Message from the Governor of Fukushima Prefecture*. Accessed June 30, 2016. http://fuku-semi.jp/iryou-pj/English/greeting.html.

Ministry of Economy, Trade and Industry (METI). 2014. Strategic Energy Plan (Provisional Translation). Tokyo: Ministry of Economy, Trade and Industry. Accessed September 18, 2015. www.enecho.meti.go.jp/en/category/others/basic_plan/pdf/4th_strategic_energy_plan.pdf.

Murakami, Yasusuke. (1999). *An Anticlassical Political–Economic Analysis: A Vision for the Next Century*. Palo Alto: Stanford University Press.

Ogawa, Hideo. 2013. "Genpatsu Jiko to Josen" [Nuclear Power Plants and Decontamination]. In *Shinsai Fukko to Jichitai: "Ningen" Fukko Eno Michi* [Post-Disaster Reconstruction and Local Governments: Road to the "Revival of the Human"], edited by Tomohiro Okada and Jichitai Mondai Kenkyusho, pp. 331–346. Tokyo: Jichitai Mondai Kenkyusho. [In Japanese]

Okada, Tomohiro. 2005. *Chiikizukuri no Keizaigaku Nyumon: Chiikinai Saitoshiryoku Ron* [Introduction to the Economics of Region-Making: Theory of Intra-Regional Reinvestment]. Tokyo: Jichitai Kenkyu Sha. [In Japanese]

Okada, Tomohiro. 2012. *Shinsai Kara no Chiiki Saisei: Ningen no Fukko ka Sanji Binjogata "Kozo Kaikaku" ka* [Regional Rebuilding from the Disaster: Human Redevelopment or "Structural Reforms": Capitalizing on Disasters]. Tokyo: Shin Nihon Shuppansha. [In Japanese]

Okada, Tomohiro, and Mitsuyoshi Kawase. 2013. *Genpatsu ni Izon Shinai Chiikizukuri Eno Tenbo: Kashiwazakishi no Chiiki Keizai to Jichitai Zaisei* [Prospects for Regions that Do Not Rely on Nuclear Power Plants: Regional Economies and Local Finance of Kashiwazaki City]. Tokyo: Jichitai Kenkyu Sha. [In Japanese]

Park, Bae-Gyoon. 2008. "Uneven Development, Inter-Scalar Tensions, and the Politics of Decentralization in South Korea." *International Journal of Urban and Regional Research* 32(1): 40–59.

Park, Bae-Gyoon, Richard Child Hill, and Asato Saito. 2012. *Locating Neoliberalism in East Asia*. West Sussex: Wiley-Blackwell.

Shimizu, Shuji. 1992. "Dengen Rittchi Sokushin Zaisei no Chiikiteki Tenkai" [Regional Implications of the Financial Scheme to Promote Power-Source Siting]. *Fukushima Daigaku Chiiki Kenkyu* 3(4): 3–26. [In Japanese]

Sone, Yasunori, Noboru Yanase, Hironobu Uekihara, and Keisuke Shimada. 2013. *Deliberative Poll*. Tokyo: Kirakusha. [In Japanese]

Suehiro, Akira. 1998. "Hatten Tojokoku no Kaihatsu Shugi" [Developmentalism in Developing Countries]. In *The 20th-Century Global System: Developmentalism* [Vol. 4], edited by Institute of Social Science, pp. 13–46. Tokyo: University of Tokyo Press. [In Japanese]

Tiebout, Charles M. 1956. "Exports and Regional Economic Growth." *Journal of Political Economy* 64(2): 160–164.

Tsukamoto, Takashi. 2011. "Neoliberalization of the Developmental State: Tokyo's Bottom-Up Politics and State Rescaling in Japan." *International Journal of Urban and Regional Research* 36(1): 71–89.

Tsukamoto, Takashi. 2012. "Why Is Japan Neoliberalizing? Rescaling of the Japanese Developmental State and Ideology of State–Capital Fixing." *Journal of Urban Affairs* 34(4): 395–418.

Woo-Cumings, Meredith. 1999. *The Developmental State*. Ithaca, NY: Cornell University Press.

Yamakawa, Mitsuo. 1987. "Posuto Dengen Kaihatsu no Ugoki – Fukushima Ken Hirono Cho no Baai" [A Movement Toward Post-Electric Power Source-based Development: The Case of Hirono Town]. *Tohoku Keizai* 81: 1–56. [In Japanese]

Yamakawa, Mitsuo. 2016. "Datsugenpatsu/Saiene Donyu to Chiiki Keizai Junkan no Kakuritsu" [Abandoning Nuclear Power, Introduction of Renewable Energy, and Establishment of Regional Economic Flows]. *Chiri* [Geography] 61(3): 60–68.

Yamakawa, Mitsuo and Daisaku Yamamoto (eds). 2016. *Unravelling the Fukushima Disaster*. London: Routledge.

Yamamoto, Daisaku, and Yumiko Yamamoto. 2013. "Community Resilience to a Developmental Shock: A Case Study of a Rural Village in Nagano, Japan." *Resilience* 1(2): 99–115.

Yamamuro, Atsushi. 1998. "Expressing the Opinion on a Controversial Issue: The Case of Nuclear Power Plant Issue in Maki Town, Niigata Prefecture." *Journal of Environmental Sociology* 4: 188–203. [In Japanese]

Index

Diagrams and photos are given in italics.